ORGANIC SYNTHESES

ORGANIC SYNTHESES

AN ANNUAL PUBLICATION OF SATISFACTORY
METHODS FOR THE PREPARATION
OF ORGANIC CHEMICALS
VOLUME 73
1996

JOHN WILEY & SONS, INC.
NEW YORK • CHICHESTER • BRISBANE • TORONTO • SINGAPORE

The procedures in this text are intended for use only by persons with prior training in the field of organic chemistry. In the checking and editing of these procedures, every effort has been made to identify potentially hazardous steps and to eliminate as much as possible the handling of potentially dangerous materials; safety precautions have been inserted where appropriate. If performed with the materials and equipment specified, in careful accordance with the instructions and methods in this text, the Editors believe the procedures to be very useful tools. However, these procedures must be conducted at one's own risk. Organic Syntheses, Inc., its Editors, who act as checkers, and its Board of Directors do not warrant or guarantee the safety of individuals using these procedures and hereby disclaim any liability for any injuries or damages claimed to have resulted from or related in any way to the procedures herein.

This text is printed on acid-free paper.

Published by John Wiley & Sons, Inc.

Library of Congress Catalog Card Number: 21-17747
ISBN 0-471-14701-X

Printed in the United States of America

10 9 8 7 6 5 4 3 2 1

ORGANIC SYNTHESES

*Out of print.
†Deceased.

v

*Out of print.
†Deceased.

Collective Volumes, Collective Indices, Annual Volumes 70–73, and Reaction Guide are available from John Wiley & Sons, Inc.

*Out of print.
†Deceased.

NOTICE

With Volume 62, the Editors of *Organic Syntheses* began a new presentation and distribution policy to shorten the time between submission and appearance of an accepted procedure. The soft cover edition of this volume is produced by a rapid and inexpensive process, and is sent at no charge to members of the Organic Divisions of the American and French Chemical Society, The Perkin Division of the Royal Society of Chemistry, and The Society of Synthetic Organic Chemistry, Japan. The soft cover edition is intended as the personal copy of the owner and is not for library use. A hard cover edition is published by John Wiley & Sons Inc. in the traditional format, and differs in content primarily in the inclusion of an index. The hard cover edition is intended primarily for library collections and is available for purchase through the publisher. Annual Volumes 65–69 have been incorporated into a new five-year version of the collective volumes of *Organic Syntheses* which have appeared as *Collective Volume Eight* in the traditional hard cover format. It is available for purchase from the publishers. The Editors hope that the new *Collective Volume* series, appearing twice as frequently as the previous decennial volumes, will provide a permanent and timely edition of the procedures for personal and institutional libraries. The Editors welcome comments and suggestions from users concerning the new editions.

NOMENCLATURE

Both common and systematic names of compounds are used throughout this volume, depending on which the Editor-in-Chief felt was more appropriate. The *Chemical Abstracts* indexing name for each title compound, if it differs from the title name, is given as a subtitle. Systematic *Chemical Abstracts* nomenclature, used in both the 9th and 10th Collective Indexes for the title compound and a selection of other compounds mentioned in the procedure, is provided in an appendix at the end of each preparation. Registry numbers, which are useful in computer searching and identification, are also provided in these appendixes. Whenever two names are concurrently in use and one name is the correct *Chemical Abstracts* name, that name is preferred.

SUBMISSION OF PREPARATIONS

Organic Syntheses welcomes and encourages submission of experimental procedures which lead to compounds of wide interest or which illustrate important new developments in methodology. The Editoral Board will consider proposals in outline format as shown below, and will request full experimental details for those proposals which are of sufficient interest. Submissions which are longer than three steps from commercial sources or from existing *Organic Syntheses* procedures will be accepted only in unusual circumstances.

Organic Syntheses Proposal Format

1) Authors
2) Title
3) Literature reference or enclose preprint if available
4) Proposed sequence
5) Best current alternative(s)
6) a. Proposed scale, final product:
   b. Overall yield:
   c. Method of isolation and purification:
   d. Purity of product (%):
   e. How determined?
7) Any unusual apparatus or experimental technique?

8) Any hazards?

9) Source of starting material?

10) Utility of method or usefulness of product

Submit to: Dr. Jeremiah P. Freeman, Secretary
Department of Chemistry
University of Notre Dame
Notre Dame, IN 46556

Proposals will be evaluated in outline form, again after submission of full experimental details and discussion, and, finally by checking experimental procedures. A form that details the preparation of a complete procedure (Notice to Submitters) may be obtained from the Secretary.

Additions, corrections, and improvements to the preparations previously published are welcomed; these should be directed to the Secretary. However, checking of such improvements will only be undertaken when new methodology is involved. Substantially improved procedures have been included in the Collective Volumes in place of a previously published procedure.

ACKNOWLEDGMENT

Organic Syntheses wishes to acknowledge the contributions of Hoffmann-La Roche, Inc. and Merck & Co. to the success of this enterprise through their support, in the form of time and expenses, of members of the Boards of Directors and Editors.

HANDLING HAZARDOUS CHEMICALS
A Brief Introduction

General Reference: *Prudent Practices for Handling Hazardous Chemicals in Laboratories,* National Academy Press, Washington, D.C. 1983

Physical Hazards

Fire. Avoid open flames by use of electric heaters. Limit the quantity of flammable liquids stored in the laboratory. Motors should be of the nonsparking induction type.

Explosion. Use shielding when working with explosive classes such as acetylides, azides, ozonides, and peroxides. Peroxidizable substances such as ethers and alkenes, when stored for a long time, should be tested for peroxides before use. Only sparkless "flammable storage" refrigerators should be used in laboratories.

Electric Shock. Use 3-prong grounded electrical equipment if possible.

Chemical Hazards

Because all chemicals are toxic under some conditions, and relatively few have been thoroughly tested, it is good strategy to minimize exposure to all chemicals. In practice this means having a good, properly installed hood; checking its performance periodically; using it properly; carrying out most operations in the hood; protecting the eyes; and, since many chemicals can penetrate the skin, avoiding skin contact by use of gloves and other protective clothing.

a. Acute Effects. These effects occur soon after exposure. The effects include burn, inflammation, allergic responses, damage to the eyes, lungs, or nervous system (e.g., dizziness), and unconsciousness or death (as from overexposure to HCN). The effect and its cause are usually obvious and so are the methods to prevent it. They generally arise from inhalation or skin contact, so should not be a problem if one follows the admonition "work in a hood and keep chemicals off your hands." Ingestion is a rare route, being generally the result of eating in the laboratory or not washing hands before eating.

b. Chronic Effects.   These effects occur after a long period of exposure or after a long latency period and may show up in any of numerous organs. Of the chronic effects of chemicals, cancer has received the most attention lately. Several dozen chemicals have been demonstrated to be carcinogenic in man and hundreds to be carcinogenic to animals. Although there is no direct correlation between carcinogenicity in animals and man, there is little doubt that a significant proportion of the chemicals used in laboratories have some potential for carcinogenicity in man. For this and other reasons, chemists should employ good practices.

The key to safe handling of chemicals is a good, properly installed hood, and the referenced book devotes many pages to hoods and ventilation. It recommends that in a laboratory where people spend much of their time working with chemicals there should be a hood for each two people, and each should have at least 2.5 linear feet (0.75 meter) of working space at it. Hoods are more than just devices to keep undesirable vapors from the laboratory atmosphere. When closed they provide a protective barrier between chemists and chemical operations, and they are a good containment device for spills. Portable shields can be a useful supplement to hoods, or can be an alternative for hazards of limited severity, e.g., for small-scale operations with oxidizing or explosive chemicals.

Specialized equipment can minimize exposure to the hazards of laboratory operations. Impact resistant safety glasses are basic equipment and should be worn at all times. They may be supplemented by face shields or goggles for particular operations, such as pouring corrosive liquids. Because skin contact with chemicals can lead to skin irritation or sensitization or, through absorption, to effects on internal organs, protective gloves are often needed.

Laboratories should have fire extinguishers and safety showers. Respirators should be available for emergencies. Emergency equipment should be kept in a central location and must be inspected periodically.

DISPOSAL OF CHEMICAL WASTE

General Reference: *Prudent Practices for Disposal of Chemicals from Laboratories*, National Academy Press, Washington, D.C. 1983

Effluents from synthetic organic chemistry fall into the following categories:

1. Gases

1a. Gaseous materials either used or generated in an organic reaction.

1b. Solvent vapors generated in reactions swept with an inert gas and during solvent stripping operations.

1c. Vapors from volatile reagents, intermediates and products.

2. Liquids

2a. Waste solvents and solvent solutions of organic solids (see item 3b).

2b. Aqueous layers from reaction work-up containing volatile organic solvents.

2c. Aqueous waste containing non-volatile organic materials.

2d. Aqueous waste containing inorganic materials.

3. Solids

3a. Metal salts and other inorganic materials.

3b. Organic residues (tars) and other unwanted organic materials.

3c. Used silica gel, charcoal, filter acids, spent catalysts and the like.

The operation of industrial scale synthetic organic chemistry in an environmentally acceptable manner* requires that all these effluent categories be dealt with properly. In small scale operations in a research or academic setting,

---

*An environmentally acceptable manner may be defined as being both in compliance with all relevant state and federal environmental regulations *and* in accord with the common sense and good judgment of an environmentally aware professional.

provision should be made for dealing with the more environmentally offensive categories.

1a. Gaseous materials that are toxic or noxious, e.g., halogens, hydrogen halides, hydrogen sulfide, ammonia, hydrogen cyanide, phosphine, nitrogen oxides, metal carbonyls, and the like.

1c. Vapors from noxious volatile organic compounds, e.g., mercaptans, sulfides, volatile amines, acrolein, acrylates, and the like.

2a. All waste solvents and solvent solutions of organic waste.

2c. Aqueous waste containing dissolved organic material known to be toxic.

2d. Aqueous waste containing dissolved inorganic material known to be toxic, particularly compounds of metals such as arsenic, beryllium, chromium, lead, manganese, mercury, nickel, and selenium.

3. All types of solid chemical waste.

Statutory procedures for waste and effluent management take precedence over any other methods. However, for operations in which compliance with statutory regulations is exempt or inapplicable because of scale or other circumstances, the following suggestions may be helpful.

Gases

Noxious gases and vapors from volatile compounds are best dealt with at the point of generation by "scrubbing" the effluent gas. The gas being swept from a reaction set-up is led through tubing to a (large!) trap to prevent suckback and on into a sintered glass gas dispersion tube immersed in the scrubbing fluid. A bleach container can be conveniently used as a vessel for the scrubbing fluid. The nature of the effluent determines which of four common fluids should be used: dilute sulfuric acid, dilute alkali or sodium carbonate solution, laundry bleach when an oxidizing scrubber is needed, and sodium thiosulfate solution or diluted alkaline sodium borohydride when a reducing scrubber is needed. Ice should be added if an exotherm is anticipated.

Larger scale operations may require the use of a pH meter or starch/iodide test paper to ensure that the scrubbing capacity is not being exceeded.

When the operation is complete, the contents of the scrubber can be poured down the laboratory sink with a large excess (10–100 volumes) of water. If the solution is a large volume of dilute acid or base, it should be neutralized before being poured down the sink.

Liquids

Every laboratory should be equipped with a waste solvent container in which *all* waste organic solvents and solutions are collected. The contents of these containers should be periodically transferred to properly labeled waste solvent drums and arrangements made for contracted disposal in a regulated and licensed incineration facility.**

Aqueous waste containing dissolved toxic organic material should be decomposed *in situ*, when feasible, by adding acid, base, oxidant, or reductant. Otherwise, the material should be concentrated to a minimum volume and added to the contents of a waste solvent drum.

Aqueous waste containing dissolved toxic inorganic material should be evaporated to dryness and the residue handled as a solid chemical waste.

Solids

Soluble organic solid waste can usually be transferred into a waste solvent drum, provided near-term incineration of the contents is assured.

Inorganic solid wastes, particularly those containing toxic metals and toxic metal compounds, used Raney nickel, manganese dioxide, etc. should be placed in glass bottles or lined fiber drums, sealed, properly labeled, and arrangements made for disposal in a secure landfill.** Used mercury is particularly pernicious and small amounts should first be amalgamated with zinc or combined with excess sulfur to solidify the material.

Other types of solid laboratory waste including used silica gel and charcoal should also be packed, labeled, and sent for disposal in a secure landfill.

Special Note

Since local ordinances may vary widely from one locale to another, one should always check with appropriate authorities. Also, professional disposal services differ in their requirements for segregating and packaging waste.

**If arrangements for incineration of waste solvent and disposal of solid chemical waste by licensed contract disposal services are not in place, a list of providers of such services should be available from a state or local office of environmental protection.

PREFACE

Annual Volume 73, as is the normal practice for this series, contains a series of 28 checked and edited procedures that describe in detail the preparation of generally useful synthetic reagents, intermediates, or products or exemplify important new synthetic methods of expected broad applicability and significance.

This collection begins with a series of three procedures illustrating important new methods for preparation of enantiomerically pure substances via asymmetric catalysis. The preparation of **3-[(1S)-1,2-DIHYDROXY-ETHYL]-1,5-DIHYDRO-3H-2,4-BENZODIOXEPINE** describes, in detail, the use of dihydroquinidine 9-O-(9'-phenanthryl) ether as a chiral ligand in the asymmetric dihydroxylation reaction which is broadly applicable for the preparation of chiral diols from monosubstituted olefins. The product, an acetal of (S)-glyceraldehyde, is itself a potentially valuable synthetic intermediate. The assembly of a chiral rhodium catalyst from methyl 2-pyrrolidone 5(R)-carboxylate and its use in the intramolecular asymmetric cyclopropanation of an allyl diazoacetate is illustrated in the preparation of **(1R,5S)-(−)-6,6-DIMETHYL-3-OXABICYCLO[3.1.0]HEXAN-2-ONE.** Another important general method for asymmetric synthesis involves the desymmetrization of bifunctional meso compounds as is described for the enantioselective enzymatic hydrolysis of cis-3,5-diacetoxycyclopentene to **(1R,4S)-(+)-4-HYDROXY-2-CYCLOPENTENYL ACETATE.** This intermediate is especially valuable as a precursor of both antipodes **(4R)-(+)-** and **(4S)-(−)-tert-BUTYLDIMETHYLSILOXY-2-CYCLOPENTEN-1-ONE,** important intermediates in the synthesis of enantiomerically pure prostanoid derivatives and other classes of natural substances, whose preparation is detailed in accompanying procedures.

Chemoselective and stereoselective general synthetic methods useful for the preparation of a variety of classes of organic molecules are illustrated by the next group of nine procedures. Stereoselective construction of trisubstituted olefins, under investigation since the late 1950's, remains a challenge. Two general, conceptually related, convergent coupling procedures are depicted in the stereodivergent preparation of **(Z)-** and **(E)-6-METHYL-6-DODECENE** via α-dimethylphenylsilyl ketones, and the highly stereoselective preparation of **(E)-2,3-DIMETHYL-3-DODECENE** via thiol esters by way of thermolysis of the derived β-lactone. A very different approach to olefin synthesis is illustrated by a procedure for preparation of **(Z)-1-ETHOXY-1-PHENYL-1-HEXENE,** which exemplifies a general preparation of trisubstituted enol ethers by alkylidenation of esters. Heteroatom directed *ortho*-metalation is a broadly applicable, highly selective, and efficient method of direct functionalization of aromatic rings. A particularly nice application of this strategy is illustrated in the preparation of **7-INDOLINECARBOXAL-DEHYDE,** which is not available via traditional aromatic substitution protocols. Development of methods providing acyclic stereocontrol have been of great interest in the past decade. One such method is exemplified by the preparation of **2,3-syn-2-METHOXYMETHOXY-1,3-NONANEDIOL** via intramolecular hydrosilylation. This procedure also details the α-metalation of enol ethers, itself a useful method for the synthesis of a variety of functionalized carbonyl derivatives. The inversion of alcohols is an important and widely employed tactic in organic synthesis. An optimized procedure employing the powerful Mitsunobu protocol is illustrated in the preparation of **(1S,2S,5R)-5-METHYL-2-(1-METHYLETHYL)CYCLOHEXYL 4-NITROBENZOATE.** Hydroboration of olefins affording anti-Markovnikov hydration of olefins is a cornerstone in the arsenal of the modern synthetic organic chemistry. Its use on large scale industrially may be limited by the need to employ hydrogen peroxide to oxidize the intermediate organoborane. The preparation of **(+)-ISOPINOCAMPHEOL** illustrates the use of sodium perborate as a safe, effective, and inexpensive alternative to the use of hydrogen peroxide for oxidation of organoboranes. A simple and effective 1,3-dihydroxylation of silyl enol ethers by oxidation with peracids is illustrated in the preparation of **16α-METHYLCORTEXOLONE** in which the corticoid sidechain is created stereospecifically in one operation. Diazocarbonyl compounds are important, versatile synthetic intermediates. The preparation of **(E)-1-DIAZO-4-PHENYL-3-BUTEN-2-ONE** illustrates a new approach employing diazotransfer to trifluoromethyl β-dicarbonyl compounds that overcomes the limitations of existing methods.

Optimized preparations of important reagents currently not readily available commercially constitute the next group of procedures. The first details an optimized preparation of a mixture of **4-DODECYLBENZENESUL-FONYL AZIDES,** which can be employed in diazotransfer chemistry such as that illustrated in the preceding procedure. The next procedure illustrates a general preparation of β-ketophosphonates exemplified by the preparation of **BIS(TRIFLUOROETHYL) (CARBOETHOXYMETHYL)PHOSPHO-NATE,** a widely used reagent for the selective preparation of (Z) electron deficient olefins via the Horner-Emmons protocol. Asymmetric hydroxylation of enols and enolates has become an increasingly important synthetic method. The last procedure in this group details the preparation of (+)-(2R,8aS)-[(8,8-DIMETHOXYCAMPHORYL)SULFONYL]OXAZIRIDINE and the related (+)-(2R,8aS)-[(8,8-DICHLOROCAMPHORYL)SULFONYL] OXAZIRIDINE, two reagents which are particularly effective for the asymmetric hydroxylation of 2-tetralones.

The next five procedures of the final set illustrate the important process of preparation of enantiomerically pure materials beginning with readily available enantiomerically pure natural substances. Use of commercially available enantiomerically pure pyrrolobenzodiazepine-5,11-diones to prepare (1S,2S)-(+)-2-(N-TOSYLAMINO)CYCLOHEXANECARBOXYLIC ACID is described in the first procedure. The next preparation illustrates an optimized preparation of DIETHYL (2S,3R)-2-(N-tert-BUTOXYCARBONYL)-AMINO-3-HYDROXYSUCCINATE from diethyl (+)-tartrate employing neighboring group participation to control substitution stereochemistry. Illustrating a method conceptually related to the Seebach methodology, the next procedure describes an optimized preparation of (R)-(−)-3-AMINO-3-(p-METHOXYPHENYL)PROPIONIC ACID from (S)-(+)-asparagine. The first of the final two procedures in this group details a general method for the preparation of γ-keto acids through metalation and alkylation of dihydrofuran followed by hydrolysis as illustrated by the procedure for preparation of 4-KETOUNDECANOIC ACID. This material serves as the starting point for an accompanying procedure which employs S-(+)-2-phenylglycinol as the chiral template for the preparation of a chiral bicyclic lactam from which S-(−)-5-HEPTYL-2-PYRROLIDINONE is derived via a series of reductions including a key stereoselective silane reduction of an acyliminium ion.

The final series of five procedures presents optimized preparations of a variety of useful organic compounds. The first procedure in this group describes the preparation of 3-BROMO-2(H)-PYRAN-2-ONE, a heterodiene useful for [4+2] cycloaddition reactions. An optimized large scale preparation of 1,3,5-CYCLOOCTATRIENE, another diene useful for [4+2] cycloaddition, is detailed from the readily available 1,5-cyclooctadiene. Previously, the availability of this material has depended on the commercial availability of cyclooctatetraene at reasonable cost. A simple large scale procedure for the preparation of 3-PYRROLINE is then presented via initial alkylation of hexamethylenetetramine with (Z)-1,4-dichloro-2-butene. This material serves as an intermediate for the preparation of 2,5-disubstituted pyrroles and pyrrolidines via heteroatom-directed metalation and alkylation of suitable derivatives. The preparation of extremely acid- and base-sensitive materials by use of the retro Diels-Alder reaction is illustrated in the preparation of **2-CYCLOHEXENE-1,4-DIONE,** a useful reactive dienophile and substrate for photochemical [2+2] cycloadditions. Functionalized ferrocene derivatives have found utility in a number of contexts, including as chiral ligands in chiral metal complexes employed in asymmetric catalysis. The next procedure details a highly optimized preparation of **ETHYNYLFERRO-CENE,** a useful intermediate for this purpose, which illustrates a general method for the transformation of aldehydes to the homologous terminal alkyne. The final procedure in this group details a fully optimized, safer method for the large scale preparation of **4,5-DIBENZOYL-1,3-DITHIOLE-1-THIONE,** an important component in the synthesis of bis(ethylenedithio)tetrathiofulvalene (BEDT-TTF), a material which forms a variety of conducting and superconducting charge-transfer salts.

The continuing success of this series derives from the dedicated efforts of many people in the Organic Syntheses family, all of whom are committed to the belief that the reliable, general procedures made available through this series serve an important function in disseminating the important new technology for modern organic synthesis not only among their professional colleagues in industry and academia around the world, but also, most importantly, to the new generation of organic chemists in training at all levels. I thank my colleagues on the Board of Editors for their assistance and dedication in the selection, checking, and in some cases, modification and improvement of the procedures presented herein. Indeed, the benefits of these efforts to our science of organic chemistry accrue in unexpected ways, as was the case when an unanticipated outcome during checking led to new insights and exciting avenues for further investigation. I am especially indebted to Professor Jeremiah P. Freeman, Secretary to the Board, and our Assistant Editor, Dr. Theodora W. Greene, whose invaluable and untiring efforts made my task and that of the entire Board of Editors most enjoyable and interesting. Dr. Greene is especially to be acknowledged for the assembly of the index.

ROBERT K. BOECKMAN, JR.

Rochester, New York
December 1994

THEODORE L. CAIRNS
July 20, 1914–September 16, 1994

Theodore L. Cairns, as a research scientist in the DuPont Company, made important contributions to the science of chemistry, applications of chemistry, and U.S. scientific policy.

Ted was born in Edmonton, Canada, in 1914. After a bachelor's degree from the University of Alberta in 1936, he earned a Ph.D. under Roger Adams at the University of Illinois in 1939. He served as an instructor in organic chemistry at the University of Rochester until 1941, when he joined the Central Research Department of DuPont in Wilmington, Delaware. There he spent the next thirty-eight years, the last eight as the Director of that department.

As a leader in the research of a major chemical company with a high reputation for innovation, his expertise and advice were sought by both scientific and governmental organizations. On the scientific side, he was a member of the National Academy of Sciences. He had an important role on several of its committees, such as those that produced the influential report ''Prudent Practices for Handling Hazardous Chemicals in the Laboratory'' and a useful survey of basic research in chemistry in the United States. He was active in the

American Chemical Society; for example, he served a term as chairman of the Division of Organic Chemistry.

Cairns was an editor of *Organic Syntheses* from 1949 to 1956, where he gained a reputation as an exceptionally diligent checker of preparations. He was an editor of its sister publication, *Organic Reactions*, from 1959 to 1969.

Cairns served on President Nixon's Scientific Advisory Committee (1970–1973) and on the Delaware Governor's Council on Science and Technology (1969–1972).

Ted's contributions to science and technology were recognized by several awards: The American Chemical Society award for creative work in synthetic organic chemistry (1968); the Perkin medal of the British Society of Chemical Industry (1973); and the Cresson Medal of the Franklin Institute (1974).

I worked directly under Ted for most of my career with the DuPont Company. Like many other DuPont chemists, I found him an encouraging and inspiring leader who took bad breaks, and even foolish blunders, in stride. These qualities served him well not only as a research director, but as an *Organic Syntheses* editor.

BLAINE C. MCKUSICK

October 21, 1994

EVAN C. HORNING
June 6, 1916–May 14, 1993

Evan C. Horning, Secretary of the Board of *Organic Syntheses* (1940–1949) and Editor-in-Chief of Collective Volume III (1995), died on May 14, 1993. After a number of years as an organic chemist, his interests changed to analytical biochemistry where he and his wife, Marjorie, made outstanding contributions to the fields of gas chromatography, mass spectrometry, and gas and liquid-mass spectrometric analysis of biological materials.

Dr. Horning was born in Philadelphia, PA, received a B.S. degree from the University of Pennsylvania (1937) and a Ph.D. from the University of Illinois (1940), doing his thesis work with Professor R. C. Fuson. He was an instructor at Bryn Mawr College during 1940–1941 and became an instructor of chemistry at the University of Michigan (1941). Here he met his wife, Marjorie, who became his co-worker and co-author of numerous publications for over 50 years.

At Michigan, he and his associates studied alkyl-substituted 2-cyclohexen-1-ones and their isomerization to aromatic compounds. Related studies were continued at the University of Pennsylvania where he became an assistant professor (1944) and an associate professor (1945–1947). His interests shifted gradually to the furans, coumarins, and morphine-type compounds. He accepted the position of Chief of the Laboratory of Chemistry at the National Institutes of Health in 1950.

Horning and his co-workers' investigations of biological materials led to methods for their purification and identification by gas chromatography. They developed procedures for separation of sugars as acetyl derivatives and steroids as trifluoroacetoxy compounds. They also were successful in using gas chromatography for studying cholesterol esters and separation of sapogenins. These investigations produced new liquid phases for gas chromatographic separations of steroids. In addition, his group investigated extensively the isolation and structure of a number of alkaloids.

From 1966 to 1986 Horning was Director of the Institute of Lipid Research, Chairman of the Biochemistry Department (1962–1966), and Professor of Chemistry at Baylor College of Medicine (1961–1986). During this period he applied gas chromatography to steroids in human fluids, urine, bile and spinal liquids, and to amino acid derivatives. His group isolated acids, alcohols, and waxes from human skin. Here they applied mass spectrometry and liquid or gas chromatography to the identification of numerous human substances. They extended these techniques to investigations of the metabolism of many different compounds in humans.

Dr. Horning received numerous honors and awards including a Guggenheim Memorial Fellowship at the University of London Post Graduate Medical School; the Torbern Bergman Award of the Swedish Society of Chemists; Dr. h.c., Karolinska Institute, Stockholm, Sweden; M.S. Tswett Medical in Chromatography; Warner-Lambert (General Diagnostics) Award, Am. Assoc. Clin. Chem.; C. W. Scheele Award, Pharmaceutical Society of Sweden; Dr. h.c., University of Ghent, Ghent, Belgium; ACS Award in Chromatography; Founders Award, Chemical Industry Institute of Toxicology; Tswett Commemorative Medal, Soviet Academy of Sciences; S.S. Dal Nogare Award in Chromatography; Southwest Science Forum Award; and the ACS Frank H. Field and Joe Franklin Award in Mass Spectrometry which he shared with his wife, Marjorie.

Horning was a member of many scientific and honorary societies and served on several editorial boards besides that of *Organic Syntheses*. They included *Journal of Medicinal Chemistry, Chemical Reviews, Analytical Biochemistry, Analytical Letters, Journal of Chromatography, Advances in Lipid.Research, Life Sciences*, and *Journal of Atherosclerosis Research*.

Evan was a quiet, soft-spoken, self-effacing man with a sly sense of humor. He was admired and enjoyed by his associates, of whom the most help of all was his devoted wife Marjorie. Not only did they share common interests at work, but enjoyed traveling, gardening, and classical music, as well as sculpture, painting, and other forms of art.

NORMAN RABJOHN

March 15, 1994

HAROLD RAY SNYDER
May 21, 1910–March 8, 1994

Harold Ray Snyder had a long and distinguished career (1937–1976) in teaching, research, editing, and administration at the University of Illinois. Steeped in the Illinois tradition, he was born at Mt. Carmel, was graduated from the University in 1931, where he did senior research with Professor R. C. Fuson, and completed his Ph.D. at Cornell University in 1935, where he did his thesis research with Professor John R. Johnson, who had previously been on the staff at Illinois. After a year at the Solvay Process Company in Syracuse, N.Y., Dr. Snyder decided to give up industrial work and to seek an academic position. Roger Adams obliged with an offer of a postdoctoral research assistantship for the academic year 1936–1937, and Harold Snyder joined the teaching staff of the University of Illinois in 1937. In that same year, Charles C. Price arrived, augmenting the "Big Four" of Adams, Marvel, Shriner, and Fuson. It is interesting, but not surprising, because Adams was one of the founders of *Organic Syntheses*, that all six of these University of Illinois staff members edited individual volumes. Harold Snyder's was volume 28. He was also on the Editorial Board of *Organic Reactions* and assumed, or perhaps was delegated, major responsibility for editing the first seven volumes during 1942–1953 while Adams was Editor-in-Chief.

Harold assembled and maintained a group of industrious, dedicated, and loyal research students that consisted first of senior undergraduates, then of a mixture of seniors and graduate students, and finally of graduate students plus an occasional postdoctorate or visiting professor on sabbatical leave. He had a close and continuing relationship with his students. This resulted in part from the hospitality, warmth, and genuine human interest of Harold and his wife, Mary Jane McIntosh Snyder. The major contribution to this relationship came from Harold's style of research direction. Some partial quotations from his former students support this conclusion. "Harold Snyder was a benevolent research supervisor who gave me as much freedom in research as I could use."—Ernest L. Eliel. "Snyder regarded research as a learning experience for his students rather than as a training exercise or a source of papers for himself."—James H. Brewster. "Well do I remember, and after all these twenty-five years recall ever more vividly, how you let me have free rein to follow my own interests in the laboratory."—Louis A. Carpino (writing to Harold in 1976). "Dr. Snyder was a tremendous teacher who always took an extra step for his thesis students and helped them throughout their careers."— Robert E. Jones.

Many students have mentioned the excellent graduate course in classical organic synthesis that Harold offered, which was also an avenue for the recruitment of some of the best research students. His attitude toward the developing mechanistic theories was that they were "useful when they suggest new experiments, but dangerous when they discourage them" (Brewster). Toward his colleagues—I write here as a one-time junior colleague—he was tolerant, generous, and helpful. He was a purist in language, so one could learn from him about style, logical development of ideas, choice of words, and grammar. He had a dry sense of humor which occasionally was capable of providing a hilarious response. Perhaps unknown to his students, but appreciated by his staff colleagues and friends, was his propensity for engaging in practical jokes, including well-staged and rather elaborate ones.

When Harold was not at his office desk he was likely to be found in the library or in the laboratory across the hall from his office in Noyes Laboratory. There he tried out new reactions on a test-tube scale before he assigned them to students, especially undergraduate research students. He explained to me that it was wise to generate a bit of optimism at the outset of a research problem. He also encouraged his graduate students to try test-tube reactions initially. His early research involved the practical synthesis of amino acids, from which logical developments followed for the synthesis of unnatural amino acids and antagonists of the natural amino acids. It is pleasing to note that his synthetic organic methodology has been applied recently to the point mutation of peptides using biotechnology.

He and his students developed C-alkylation with quaternary ammonium salts and nucleophilic displacements on such salts, including the stereochemistry. His name is immediately associated with important innovations in the use of polyphosphoric acid for inter- and intramolecular condensations, cyclizations, and functional conversions in organic chemistry. He pioneered the use of boron trifluoride as an efficient catalyst in the Fischer indole synthesis and discovered new reactions of anils, including Diels-Alder reactions. He and his students delineated the requirements for disproportionation of tertiary amines. He developed the synthesis and chemistry of arylboronic acids. One of his fundamental ideas was for the incorporation of sufficient boron into organ-specific drugs that they could then be selectively neutron-irradiated at their *in vivo* locus. Chemists/pharmacologists are still trying to meet this challenge. In addition to Harold's many research publications and patents, he coauthored a textbook, *Organic Chemistry* (John Wiley, 1st edition 1942) with R. C. Fuson.

During the Second World War, Harold carried out work for the National Defense Research Committee, the War Production Board Rubber Research Program (with Marvel), and the Committee on Medical Research. Within the latter program, the team of Snyder, Price, and Leonard, together with their graduate students, helped develop the process for production of the antimalarial drug Chloroquine in time for its use in the Pacific. The drug is still in use today, although resistant strains of parasites are a problem in some areas of the world. Just before the war, Harold had been awarded a John Simon Guggenheim fellowship (1939), which represented an unusually early appreciation of his record in chemical research. However, his trip to Europe, which included a series of lectures in Italy, had to be postponed until 1952. A Professor of Chemistry at Illinois from 1945, Harold served as Associate Head of the Department during 1957–1960 and Associate Dean of the Graduate College during 1960–1975. In the latter role, he also served as Secretary of the Research Board, which was a very responsible position because the Board distributed all internal research grants and graduate fellowships. It is clear that Harold's objectivity and his modest nature contributed to his long and effective tenure in that office. Outside the University, Harold was a consultant to Merck and Company and to Phillips Petroleum Company. He was also an adviser to the Office of Naval Research in the evaluation of ONR research grant applications.

A symposium in Honor of Harold Ray Snyder was held in 1976 which was attended by friends, associates, and former students who gathered to pay tribute. The Symposium also marked his 65th year and his retirement from the University of Illinois. Further, his former students organized the Harold Snyder Endowment Fund, which supports undergraduate students, who have an interest in organic chemistry, to do research with a University of Illinois faculty member. Since 1990, two Snyder Scholars have been supported annually. Additional scholarships are probable with increases in the Endowment Fund.

Harold Snyder was predeceased by his first wife, Mary Jane Snyder, and by his second wife, her sister Bonnie McIntosh Snyder. He is survived by his three children, Dr. Jane Snyder of Columbus, Ohio, Dr. John Snyder of Basking Ridge, New Jersey, and Dr. Mary Ann Nirdlinger of Sylvania, Ohio; four grandchildren and several nieces and nephews; and a sister, Joanne Dorsch of Lake Kiowa, Texas. He will be long remembered by his own family and by the many former students and colleagues who constitute his professional family.

NELSON J. LEONARD

June 16, 1994

CONTENTS

ORGANIC SYNTHESES

3-[(1S)-1,2-DIHYDROXYETHYL]-1,5-DIHYDRO-3H-2,4-BENZODIOXEPINE

(1,2-Ethanediol, 1-(1,5-dihydro-2,4-benzodioxepin-3-yl)-, (S)-)

A.

B.

C.

Submitted by Ryu Oi[1] and K. Barry Sharpless.[2]

Checked by Michelle A. Laci and Robert K. Boeckman, Jr.

Caution: Lithium aluminum hydride is a flammable solid which evolves hydrogen, a highly flammable and potentially explosive gas, upon contact with moisture. Potassium osmate is highly toxic as are any residues containing osmium and should be handled with care. All operations should be conducted in an efficient fume hood.

1

1. Procedure

A. 1,2-Benzenedimethanol . An oven-dried, 2-L, three-necked, round-bottomed flask, equipped for mechanical stirring, and outfitted with a reflux condenser bearing an argon inlet/outlet vented through a mineral oil bubbler, and a 250-mL, pressure-equalizing addition funnel, is placed under an argon atmosphere and charged with 20.94 g (0.552 mol) of lithium aluminum hydride ($LiAlH_4$), and 850 mL of anhydrous diethyl ether (Notes 1 and 2). A solution of 87.46 mL (97.78 g, 0.44 mol) of diethyl phthalate (Note 1) in 170 mL of anhydrous tetrahydrofuran (THF) is added dropwise via the addition funnel over 1.5 hr with efficient mechanical stirring, at a rate of addition adjusted to maintain a gentle reflux (Note 3). After the addition is complete, the mixture is heated to reflux by means of a heating mantle for 1.5 hr. The mechanically stirred reaction mixture is then cooled to ~0°C in an ice-water bath, and quenched by *cautious* sequential addition of 21 mL of water, 21 mL of aqueous 15% sodium hydroxide solution, and 63 mL of water (Note 4). While stirring is continued, the mixture is held for 30 min at 0°C then allowed to warm to room temperature. The resulting white solids are collected by vacuum filtration and washed on the filter using a total of 1 L of ether in 100-mL portions (Note 5). Concentration of the filtrate, under reduced pressure, affords 56.64 g (93%) of 1,2-benzenedimethanol as a white solid, mp 62-64°C, which is sufficiently pure for further transformation (Note 6).

B.   3-Vinyl-1,5-dihydro-3H-2,4-benzodioxepine. An oven-dried, 2-L, three-necked, round-bottomed flask is equipped with a magnetic stirring bar, a calcium chloride drying tube, and two rubber septa. The flask is charged with 41.45 g (0.3 mol) of 1,2-benzenedimethanol, 5.71 g (0.03 mol) of p-toluenesulfonic acid monohydrate (Note 1), and 600 mL of dry 1,2-dimethoxyethane (Notes 1 and 7). Stirring is initiated, and 49.2 mL (0.45 mol) of trimethyl orthoformate (Note 1) and 30.1 mL (0.45 mol) of acrolein (Note 1) are then added sequentially via syringe. The reaction mixture is

2

stirred at room temperature for 5 hr (Note 8), then poured into a mixture of ice and 150 mL of aqueous 10% sodium bicarbonate solution in a 2-L separatory funnel using 90 mL of ether to rinse the flask. After thorough mixing, the organic phase is separated and the aqueous phase is extracted with 300 mL of ether. The combined organic layers are washed sequentially with 90 mL of cold aqueous 10% sodium bicarbonate solution and 90 mL of brine, dried over anhydrous sodium sulfate (Na_2SO_4), filtered, and concentrated under reduced pressure to give 52 g (99%) of crude acetal. Distillation of the residual liquid under reduced pressure affords 39.6-40.4 g (75-77%) of 3-vinyl-1,5-dihydro-3H-2,4-benzodioxepine as a colorless liquid, bp 116-118°C (5 mm), that may crystallize on standing, mp ~30°C (Notes 9 and 10).

    C.   3-[(1S)-1,2-Dihydroxyethyl)]-1,5-dihydro-3H-2,4-benzodioxepine. A 2-L, three-necked, round-bottomed flask is equipped with a mechanical stirrer, a thermometer, and a rubber septum. The flask is charged with dihydroquinidine 9-O-(9'-phenanthryl) ether (Notes 1 and 11), [0.75 g, (1.5 mmol, 1 mol%, 0.0057 M in the organic phase)], 148.2 g (0.45 mol) of potassium ferricyanide (Note 1), 62.2 g (0.45 mol) of potassium carbonate (Note 1), 264 mL of tert-butyl alcohol (tert-BuOH) (Note 1), and 264 mL of distilled water. Mechanical stirring is initiated, 0.11 g (0.3 mmol, 0.2 mol%) of potassium osmate (VI) dihydrate, which generates a 0.0011 M solution of OsO_4 in the organic phase (Note 1), is added, and the mixture is stirred at room temperature for 0.5 hr, resulting in an orange suspension. The flask is immersed in a 0°C cooling bath, and the mixture stirred for 1 hr (Note 12). Over a period of 24 hr at 0°C, 26.25 g (0.16 mol) of 3-vinyl-1,5-dihydro-3H-2,4-benzodioxepine is added dropwise via syringe (Note 13), and the reaction mixture is stirred an additional 24 hr at 0°C (Note 14). After the required time period has elapsed, 30.25 g (0.24 mol) of solid sodium sulfite (Na_2SO_3) is added in portions, the reaction mixture is stirred for an additional 1 hr at room temperature, poured into a 3-L beaker, and allowed to stand for another 1 hr at room temperature (Note 1). After the organic phase is decanted, the

3

aqueous phase is diluted with 300 mL of water to dissolve the salts, and poured into a 2-L separatory funnel using 90 mL of dichloromethane (CH_2Cl_2) to rinse the flask and beaker. The aqueous phase is extracted three times with 150 mL of CH_2Cl_2. The organic extracts are combined with the decanted organic phase, dried over Na_2SO_4, filtered, and concentrated, under reduced pressure, to afford 33.1 g (104%) of crude product, as a pale yellow liquid (Note 15) that crystallizes rapidly. The ee of the crude 3-[(1S)-1,2-dihydroxyethyl)]-1,5-dihydro-3H-2,4-benzodioxepine is determined by HPLC analysis of the derived bis-Mosher ester to be 84% (Note 16).

A 2-L Erlenmeyer flask is charged with 33.1 g of the crude protected diol which is dissolved in 600 mL of hot ethyl acetate (EtOAc), and the solution is allowed to stand at room temperature for 3 hr and at 0°C to -5°C for 24 hr (Note 17). The resulting white precipitate is collected by suction filtration through a 10-cm Büchner funnel (Note 18), and the mother liquor is concentrated under reduced pressure to afford 21.15 g of a pale yellow liquid that slowly crystallizes (Note 19). The residual liquid is placed in a 500-mL Erlenmeyer flask, dissolved in 150 mL of hot toluene (Note 20), and allowed to stand for 24 hr at 0°C to -5°C. The slightly yellow precipitate is collected by suction filtration through a 10-cm Büchner funnel and dried under reduced pressure to give 16.0-17.4 g (50-55%) of 3-[(1S)-1,2-dihydroxyethyl)]-1,5-dihydro-3H-2,4-benzo-dioxepine, mp 73-75°C, $[\alpha]_D^{23}$ -12.4° ($CHCl_3$, c 2.62) (Note 21). The ee of the product is determined by HPLC analysis of the derived bis-Mosher ester to be 97% (Note 16).

The mother liquor is then concentrated to afford 2.4 g of a brown yellow liquid (Note 22). Flash chromatography of this material on 30 g of SiO_2 with elution by EtOAc then by EtOAc:MeOH (5:1 v/v) affords 0.63 g (2%) of 3-[(1S)-1,2-dihydroxyethyl)]-1,5-dihydro-3H-2,4-benzodioxepine and 0.69 g (92% recovery) of dihydroquinidine 9-O-(9'-phenanthryl) ether.

2. Notes

1. This reagent was purchased from Aldrich Chemical Company, Inc., and used without further purification.

2. Anhydrous ether was purchased from J. T. Baker Chemical Company and used as received.

3. Tetrahydrofuran was purchased from J. T. Baker Chemical Company and dried and deoxygenated prior to use by distillation from sodium-benzophenone ketyl under argon.

4. During the initial stages of the addition of water, care must be exercised to control the vigorous exothermic reaction by use of efficient cooling and agitation along with careful regulation of the rate of addition. This procedure must be conducted in an efficient fume hood since a considerable volume of hydrogen gas is produced that must be vented.

5. Since the solids have a tendency to occlude significant amounts of the diol, efficient washing of the solids is essential to obtain good yields. Warm solvent can be employed to assist in removal of the product.

6. 1,2-Benzenedimethanol (lit.[3] mp 63-65°C) has the following spectral characteristics: ^{1}H NMR (300 MHz, CDCl$_3$) δ: 3.97 (s, 2 H), 4.61 (s, 4 H), 7.30 (s, 4 H); ^{13}C NMR (75 MHz, CDCl$_3$) δ: 63.8, 128.5, 129.6, 139.3; IR (CHCl$_3$) cm^{-1}: 3279, 2890, 1454, 1214, 1182, 1110, 1036, 1005, 758.

7. A freshly opened bottle of reagent grade 1,2-dimethoxyethane (Aldrich Chemical Company, Inc.) was used without further purification.

8. The reaction mixture was monitored by TLC analysis on silica gel (4:1 EtOAc:hexane v/v). 3-Methoxy-1,5-dihydro-3H-2,4-benzodioxepine (R_f = 0.5) is first produced by transesterification with trimethyl orthoformate, and is then transformed to the final product (R_f = 0.6).

9. Slightly impure forerun, 6-9 g, was separated from the main fraction. After the distillation was stopped, approximately 6-9 g of residue remained.

10. The spectroscopic data and elemental analysis of the product are as follows: ^{1}H NMR (300 MHz, CDCl$_3$) δ: 4.91 (q, 4 H, J = 14.2), 5.33-5.38 (m, 2 H), 5.53 (dt, 1 H, J = 17.4, 1.4), 5.90-5.99 (m, 1 H), 7.10-7.25 (m, 4 H); ^{13}C NMR (75 MHz, CDCl$_3$) δ: 70.0, 104.2, 118.4, 127.0, 127.2, 134.7, 138.8; IR (neat) cm^{-1}: 3068, 3024, 2950, 2860, 1769, 1725, 1497, 1445, 1412, 1376, 1353, 1291, 1268, 1221, 1210, 1150, 1123, 1094, 1046, 1032, 944, 932, 749; m/z Calcd. for Cs$^+$-C$_{11}$H$_{12}$O$_2$: 308.9892; Found 308.9892; Anal. Calcd for C$_{11}$H$_{12}$O$_2$: C, 74.98; H, 6.86; O, 18.16. Found: C, 75.12; H, 6.84; O, 18.26.

11. Dihydroquinidine 9-O-(9'-phenanthryl) ether[4] (Aldrich, Cat. No. 38195-0: Hydroquinidine 9-phenanthryl ether) gave the highest enantioselectivity compared with several other commercially available dihydroquinidine derivatives; 34% ee with dihydroquinidine 4-chlorobenzoate (Aldrich, Cat. No. 33648-3: Hydroquinidine 4-chlorobenzoate) and 61% ee with dihydroquinidine-9-O-(4'-methyl-2'-quinolyl) ether (Aldrich, Cat. No. 38194-2: Hydroquinidine 4-methyl-2-quinolyl ether). The submitters report that the reaction can be run on 0.5-mol scale using this procedure to afford comparable yields and enantiomeric purity of the diol product.

12. Upon cooling, the mixture becomes a viscous suspension, and vigorous stirring is required. At temperatures below -5°C, the reaction mixture starts to freeze, thereby precluding efficient mixing.

13. Neat 3-vinyl-1,5-dihydro-3H-2,4-benzodioxepine was added by syringe drive (Syringe Infusion Pump 22, Harvard Apparatus, South Natick, MA) using a gas-tight syringe. If the material has crystallized, it is liquified by gentle warming (~30°C) and 2 mL of t-BuOH-H$_2$O is added before placing the material in the syringe. The tip of the syringe needle is 15 cm above the reaction surface to prevent freezing of the liquid which occurs around 5°C. Careful temperature control is required to prevent the

mixture from freezing during the slow addition. Without slow addition, the ee of the product drops to 67%. The ee also deteriorates if the mixture is allowed to freeze during addition, permitting unreacted olefin to accumulate prior to warming enough to achieve efficient stirring.

14. Reaction progress was monitored by TLC on silica gel (5:1 EtOAc:MeOH v/v). As the reaction neared completion, the color of the reaction mixture turned yellow.

15. The oil contained 5-10% t-BuOH by weight. 3-[(1R)-1,2-Dihydroxyethyl)]-1,5-dihydro-3H-2,4-benzodioxepine was produced with 84% ee when dihydroquinine-9-O-(9'-phenanthryl) ether (Aldrich, Cat. No. 38197-7: Hydroquinine 9-phenanthryl ether)' (Note 1) was used instead of dihydroquinidine-9-O-(9'-phenanthryl) ether.[5]

16. The ee was determined by HPLC analysis (Chemcosorb 3Si, Chemco, Japan) of the corresponding bis-Mosher[6] ester eluted with 5% EtOAc in hexane (2 mL/min, (R)-diol: t_R = 14.8 min, (S)-diol: t_R = 18.2 min). The checkers employed a 25 x 10-mm Prep Nova-Pak HR Silica Column (particle size 6 μm, 60 Å) with UV detection at 220 nm and elution by 10% diethyl ether/hexane at 12 mL per min flow rate (R-diol: t_R = 12.6 min and S-diol: t_R = 15.1 min).

17. Ethyl acetate is preferable to aromatic hydrocarbon solvents such as benzene or toluene because of the high solubility of the diol in EtOAc; this results in higher recovery of the enantiomerically enriched diol from the mother liquor.

18. The white precipitate was dried under reduced pressure to afford 8.7-11.1 g (28-35%) of 3-[(1S)-1,2-dihydroxyethyl)]-1,5-dihydro-3H-2,4-benzodioxepine (55-60% ee).

19. The ee of this crude mixture before recrystallization was 97%.

20. Aromatic hydrocarbon solvents such as toluene or benzene are preferable to several other solvents such as EtOAc or ethanol because of efficient recovery of the diol and better separation of the diol from the ligand.

7

21. The melting point of the racemic diol was 110-113°C. The spectral data and elemental analysis of the diol are as follows: ^{1}H NMR (300 MHz, CDCl$_3$) δ: 2.13 (t, 1 H, J = 6.4), 2.57 (d, 1 H, J = 4.2), 3.73 (m, 1 H), 3.80 (dd, 2 H, J = 6.4, 4.2), 4.93-5.01 (m, 5 H), 7.20-7 30 (m, 4 H); ^{13}C NMR (75 MHz, CDCl$_3$) δ: 62.42, 72.17, 72.86, 73.23, 108.20, 127.73, 127.79, 138.89, 138.93; IR (CHCl$_3$) cm^{-1}: 3575, 3018, 2966, 2892, 2855, 1456, 1446, 1375, 1294, 1265, 1138, 1103, 1083, 1048; m/z Calcd. for Na$^+$-C$_{11}$H$_{14}$O$_4$: 233.0790; Found 233.0970; Anal. Calcd for C$_{11}$H$_{14}$O$_4$: C, 62.85; H, 6.71; Found: C, 62.78; H, 6.57.

22. The oil contained 10-30% toluene by weight.

Waste Disposal Information

All toxic materials were disposed of in accordance with "Prudent Practices for Disposal of Chemicals from Laboratories"; National Academic Press; Washington, DC, 1983.

3. Discussion

This procedure describes a convenient preparation of highly enantiomerically enriched 3-(1,2-dihydroxyethyl)-1,5-dihydro-3H,2,4-benzodioxepine,[5] a chiral glyceraldehyde equivalent, on a 0.5-mol scale and illustrates the utility of catalytic asymmetric dihydroxylation (ADH)[7] on a large scale. The merits of the diol compared with other chiral derivatives of glyceraldehyde, e.g., 2,3-O-isopropylidene-, 2,3-O-cyclohexylidene-, or 2,3-di-O-acylated glyceraldehyde[8] include its ease of handling, its low volatility and high stability at ambient temperature, and its UV chromophore that facilitates TLC or HPLC analysis. In addition, this seven-membered ring acetal can be deprotected easily under mild, neutral conditions by catalytic hydrogenolysis.[5,9] A variety of selective nucleophilic substitution reactions on derivatives easily obtained from this new C-3 chiral building block can be envisioned.[5]

The ADH at higher substrate concentrations is more practical for large scale applications. Unfortunately, high concentration of olefin in the reaction mixture is detrimental to the catalytic cycle responsible for asymmetric induction as the ee dropped to 67% at high olefin concentration (0.57 M in t-BuOH) compared with 86% at low concentrations (0.13 M in t-BuOH). However, in the more concentrated case, slow addition of the olefin raised the ee from 67% to 84%. The mechanistic rationale for the deleterious effect of high olefin concentration in this $Os/K_3Fe(CN)_6$ system[10] is not yet clear, but in any case the slow addition used here keeps the olefin concentration low.

Of the three types of racemates,[11] conglomerate, racemic compound, and solid solution, 3-(1,2-dihydroxyethyl)-1,5-dihydro-3H-2,4-benzodioxepine shows melting point behavior characteristic of a racemic compound. The racemic diol is much higher melting than the enantiomerically enriched diol as shown in the Figure 1. Therefore the diol of lower ee precipitates first during recrystallization and the enantiomerically enriched diol remains in the mother liquor. High ee diol (97% ee) is then obtained upon recrystallization of this mother liquor.

Figure 1

Relationship Between Enantiopurity and Melting Point

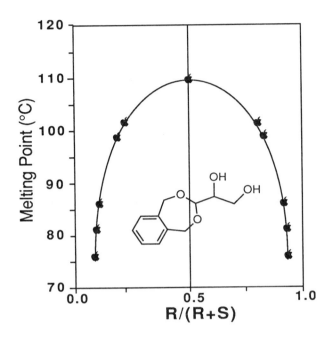

1. On leave from Omuta Research Institute, Mitsui Toatsu Chemicals, Inc., Omuta, Fukuoka 836, Japan.

2. Department of Chemistry, The Scripps Research Institute, La Jolla, CA 92037.

3. Soai, K.; Oyamada, H.; Takase, M.; Ookawa, A. *Bull. Chem. Soc. Jpn.* **1984**, *57*, 1948.

4. Sharpless, K. B.; Amberg, W.; Beller, M.; Chen, H.; Hartung, J.; Kawanami, Y.; Lübben, D.; Manoury, E.; Ogino, Y.; Shibata, T.; Ukita, T. *J. Org. Chem.* **1991**, *56*, 4585.

5. Oi, R.; Sharpless, K. B. *Tetrahedron Lett.*, **1992**, *33*, 2095.

6. Dale, J. A.; Mosher, H. S. *J. Am. Chem. Soc.* **1973**, *95*, 512.

7. Recently a new ligand was introduced for terminal olefins with branching on the pendant substituent (Crispino, G.; Jeong, K.-S.; Kolb, H. C.; Wang, Z.-M.; Xu, D.; Sharpless, K. B. *J. Org. Chem.* **1993**, *58*, 3785), but the phenanthryl ether ligand is on rare occasions still the best ligand for certain types of terminal olefins (the present acrolein acetal substrate is such a case).

8. (a) For a review of chiral glyceraldehyde equivalents, see: Jurczak, J.; Pikul, S.; Bauer, T. *Tetrahedron* **1986**, *42*, 447; (b) For a review of applications of chiral glycidol and its related compounds, see: Hanson, R. M. *Chem. Rev.* **1991**, *91*, 437.

9. Acetals of 1,2-benzenedimethanol have been prepared previously and used for carbonyl protection; (a) Machinaga, N.; Kibayashi, C. *Tetrahedron Lett.* **1989**, *30*, 4165; (b) Burke, S. D.; Deaton, D. N. *Tetrahedron Lett.* **1991**, *32*, 4651, and references therein.

10. (a) Kwong, H.-L.; Sorato, C.; Ogino, Y.; Chen, H.; Sharpless, K. B. *Tetrahedron Lett.* **1990**, *31*, 2999; (b) Ogino, Y.; Chen, H.; Kwong, H.-L.; Sharpless, K. B. *Tetrahedron Lett.* **1991**, *32*, 3965.

11. Jacques, J.; Collet, A.; Wilen, S. H., Ed., "Enantiomers, Racemates, and Resolution"; John Wiley & Sons: New York, 1981.

Appendix
Chemical Abstracts Nomenclature (Collective Index Number); (Registry Number)

3-[(1S)-1,2-Dihydroxyethyl)]-1,5-dihydro-3H-2,4-benzodioxepine: 1,2-Ethanediol, 1-(1,5-dihydro-2,4-benzodioxepin-3-yl)-, (S)-(13); (142235-22-1)

1,2-Benzenedimethanol (9); (612-14-6)

Lithium aluminum hydride: Aluminate (1-), tetrahydro-, lithium (8); Aluminate (1-), tetrahydro-, lithium (T-4)- (9); (16853-85-3)

Diethyl phthalate: Phthalic acid, diethyl ester (8); 1,2-Benzenedicarboxylic acid, diethyl ester (9); (84-66-2)

3-Vinyl-1,5-dihydro-3H-2,4-benzodioxepine: 2,4-Benzodioxepin, 3-ethyl-1,5-dihydro- (13); (142169-23-1)

p-Toluenesulfonic acid monohydrate (8); Benzenesulfonic acid, 4-methyl-, monohydrate (9); (6192-52-5)

Dimethoxyethane: Ethane, 1,2-dimethoxy- (8,9); (110-71-4)

Trimethyl orthoformate: Orthoformic acid, trimethyl ester (8); Methane, trimethoxy- (9); (149-73-5)

Acrolein (8); 2-Propenal (9); (107-02-8)

Dihydroquinidine 9-O-(9'-phenanthryl) ether: Cinchonan, 10,11-dihydro-6'-methoxy-9 (9-phenanthrenyloxy)-, (9S)-(13); (135042-88-5)

Potassium ferricyanide: Ferrate (3-), hexacyano-, tripotassium (8); Ferrate (3-), hexakis(cyano-C)-, tripotassium, (OC-6-11)- (9); (13746-66-2)

Potassium osmate(VI) dihydrate: Osmic acid, dipotassium salt (8,9); (19718-36-3)

(1R,5S)-(-)-6,6-DIMETHYL-3-OXABICYCLO[3.1.0]HEXAN-2-ONE. HIGHLY ENANTIOSELECTIVE INTRAMOLECULAR CYCLOPROPANATION CATALYZED BY DIRHODIUM(II) TETRAKIS[METHYL 2-PYRROLIDONE-5(R)-CARBOXYLATE]

((3-Oxabicyclo[3.1.0]hexan-2-one, 6,6-dimethyl-, (1R-cis)-)

Submitted by Michael P. Doyle, William R. Winchester, Marina N. Protopopova, Amy P. Kazala, and Larry J. Westrum.[1]

Checked by Robert K. Boeckman, Jr., Jacqueline C. Bussolari, and Michael R. Reeder.

1. Procedure

Caution! Thionyl chloride is a reactive substance that must be handled in a fume hood.

  A. *(-)-Methyl 2-pyrrolidone-5(R)-carboxylate.* A 500-mL, single-necked, round-bottomed flask, equipped with a Teflon-coated magnetic stirring bar, loose fitting rubber septum, and a needle for nitrogen inlet which is vented through a mineral oil bubbler, is flushed with nitrogen and charged with 6.60 g (51.1 mmol) of (R)-(+)-2-pyrrolidone-5-carboxylic acid (Note 1) and 200 mL of methanol (Note 2). Thionyl chloride (0.50 mL, 6.8 mmol) is added cautiously by syringe over a 0.5-min period to the rapidly stirring solution at room temperature. The reaction flask is sealed with the septum and the nitrogen inlet removed after addition is complete, and the resulting solution is stirred continuously at room temperature for 24 hr. The flask is uncapped, and aqueous, saturated sodium carbonate is added with stirring until the reaction solution is pH 7 to litmus paper (Note 3). After removal of the methanol at room temperature under reduced pressure by rotary evaporator, the residue is dissolved in 50 mL of dichloromethane and washed with 10 mL of aqueous, saturated sodium chloride solution, dried over anhydrous magnesium sulfate, filtered, and concentrated under reduced pressure. The residue is distilled through a short-path distillation apparatus to afford 5.60 g (75% yield) of colorless methyl (R)-2-pyrrolidone-5-carboxylate, bp 83°C (0.25 mm), $[\alpha]_D^{24}$ -8.6° (EtOH, *c* 1.13) (Note 4).

  B. *(+)-Dirhodium(II) tetrakis[methyl 2-pyrrolidone-5(R)-carboxylate] acetonitrile-isopropanol solvate,* (+)-Rh₂[(5R)-MEPY]₄(CH₃CN)₂(*i*-PrOH). A 50-mL, single-necked, round-bottomed flask, equipped with a Teflon-coated stirring bar and nitrogen inlet, is flushed with nitrogen and charged with 200 mg (0.452 mmol) of rhodium(II) acetate (Note 5), 1.02 g (7.13 mmol) of methyl 2-pyrrolidone-5(R)-carboxylate, and 20 mL of chlorobenzene (Note 6). The flask is fitted with a Soxhlet extraction apparatus (Note 7) into which is placed a thimble containing 5 g of an oven-dried mixture of 2 parts sodium carbonate and 1 part of sand (Note 8). The initially green rhodium acetate forms a blue solution upon mixing with the pyrrolidonecarboxylate which is heated at reflux under nitrogen for 6 hr (Note 9) followed by removal of the solvent under reduced pressure using a rotary evaporator to yield a blue, glass-like solid. The solid is dissolved in a minimal volume of methanol and purified by chromatography on a column containing J.T. Baker BAKERBOND Cyano 40 μm prep LC pacKing (10 g) in methanol. The dirhodium(II) product forms a broad blue band across the top of the column, which moves imperceptibly with methanol. However, the excess ligand moves rapidly as a yellow/brown band in methanol and may be collected with the first 100 mL of eluent (Note 10). Subsequent elution with 1.0 vol % acetonitrile in methanol causes an instantaneous color change from blue to red and rapid elution. The entire red band is collected as one fraction, and the solvent is removed under reduced pressure (Notes 11, 12). The resulting blue, glass-like solid (380 mg) is recrystallized by dissolving the solid in 1.0 mL of acetonitrile per 100 mg of solid and then adding an equivalent amount of isopropyl alcohol. Bright red crystals of $Rh_2(5R-MEPY)_4(CH_3CN)_2(i\text{-PrOH})$ form overnight, and are collected by suction, washed with isopropyl alcohol and air-dried to yield 240 mg (0.262 mmol, 58% yield) (Note 13).

C. *3-Methyl-2-buten-1-yl acetoacetate.* A 500-mL, three-necked, round-bottomed flask, equipped with a Teflon-coated magnetic stirring bar, stoppered pressure-equalizing addition funnel and a reflux condenser bearing a drying tube, is charged with a solution of 21.5 g (0.250 mol) of 3-methyl-2-buten-1-ol in 70 mL of anhydrous tetrahydrofuran and 1.20 g (0.015 mol) of sodium acetate (Note 14). The reaction mixture is heated at reflux with continuous magnetic stirring, and a solution of 23.5 g (0.280 mol) of freshly distilled diketene in 30 mL of tetrahydrofuran is added dropwise *via* the addition funnel over 1 hr (Note 15). The reaction mixture is heated at reflux for an additional 30 min upon completion of the addition. The brown reaction mixture is cooled to room temperature, transferred to a separatory funnel containing 300 mL of ether and 50 mL of saturated aqueous sodium chloride solution, mixed thoroughly, and the aqueous layer withdrawn. The aqueous layer is extracted twice with 50-mL portions of ether, and the combined ether solution is washed twice with 50-mL portions of saturated aqueous sodium chloride solution. After the organic layer is dried over anhydrous magnesium sulfate, the solvent is removed under reduced pressure, and the brown residue is distilled (bp 85°C, 0.15 mm) to afford 28.9 g (68% yield)of 3-methyl-2-buten-1-yl acetoacetate as a colorless liquid (Note 16).

D.   3-Methyl-2-buten-1-yl diazoacetate.   A 1-L, three-necked, round-bottomed flask, equipped with a Teflon-coated magnetic stirring bar and a stoppered, pressure-equalizing addition funnel is charged with a solution of 30.5 g (0.179 mmol) of 3-methyl-2-buten-1-yl acetoacetate and 23.3 g (0.231 mol) of triethylamine in 150 mL of anhydrous acetonitrile (Note 16).   A solution of 55.85 g (0.233 mol) p-acetamido-benzenesulfonyl azide in 150 mL of acetonitrile is added dropwise over 30 min with continuous magnetic stirring (Note 17).   A white precipitate of p-acetamidobenzene-sulfonamide is observed after ~30 min (additional acetonitrile is added if necessary to facilitate stirring). The resulting heterogeneous yellow mixture is maintained at room temperature for an additional 2.5-3 hr (Note 18). A solution of 24.6 g (0.586 mol) of lithium hydroxide (LiOH·H2O) in 200 mL of water is added to the reaction mixture (the precipitate dissolves), and the resulting light brown mixture is stirred for 8 hr (Note 19). The reaction mixture is extracted with two 50-mL portions of 2:1 ether:ethyl acetate. The combined organic phases are washed with one 25-mL portion of saturated aqueous sodium chloride solution, dried over anhydrous magnesium sulfate, and the solvent is removed under reduced pressure. The resulting orange liquid, containing some solid sulfonamide, is purified by flash column chromatography on silica (200 g) with 1:10 ethyl acetate:hexane as the eluent. Collection of the yellow band and concentration under reduced pressure affords 22.39 g (81% yield) of 3-methyl-2-butenyl diazoacetate as a yellow oil (Note 20).

    E.   *(1R,5S)-(-)-6,6-Dimethyl-3-oxabicyclo[3.1.0]hexan-2-one.* A 1-L, three-necked, round-bottomed flask, equipped with a stoppered pressure-equalizing addition funnel and a reflux condenser bearing a drying tube and a nitrogen inlet vented through a mineral oil bubbler, is charged with a solution of 0.203 g (0.221 mmol) of $Rh_2(5R-MEPY)_4(CH_3CN)_2(i-PrOH)$ in 150 mL of freshly distilled, anhydrous dichloromethane (Note 21). The apparatus is flushed with nitrogen and the reaction mixture is heated at reflux during the dropwise addition of a solution of 14.9 g (96.7 mmol) of 3-methyl-2-buten-1-yl diazoacetate in 450 mL of anhydrous dichloromethane over a 30-hr period. The slow flow of external inert gas is terminated after 1 hr, and the evolution of nitrogen from the reaction solution is used to monitor the progress of the reaction. After the addition is complete, the blue reaction mixture is heated at reflux for an additional hour, cooled to room temperature, and the solvent is removed from the green reaction mixture under reduced pressure. Kugelrohr distillation of the residue (80°C, 0.15 mm) affords 12.1 g of a nearly colorless liquid that upon chromatographic purification on silica-alumina (100 g) with hexane:ethyl acetate (80:20 to 70:30) affords 10.2-10.7 g (84-88% yield) of >99% pure (1R,5S)-(-)-6,6-dimethyl-2-oxabicyclo-[3.1.0]hexan-2-one, bp 70°C (0.15 mm), $[\alpha]_D^{24}$ -81.2° (CHCl$_3$, *c* 2.4) (Note 22), whose optical purity was determined on a Chiraldex B-PH capillary column to be 92-93% ee (Note 23). The solid residue remaining after Kugelrohr distillation is dissolved in a minimal volume of methanol and purified by chromatography, as described in Part B. Isolation of the red band affords 0.098 - 0.099g (~49%) of recovered catalyst (Note 24).

2. Notes

1. (R)-(+)-2-Pyrrolidone-5-carboxylic acid [Fluka Chemika-BioChemika, D-pyroglutamic acid, $[\alpha]_D^{22}$ +9±1° (H$_2$O, c 5)] was recrystallized from ethanol/ether (2/1), $[\alpha]_D^{22}$ +10° (H$_2$O, c 0.97), prior to use.

2. HPLC grade methanol (Aldrich Chemical Company, Inc.) was used without further purification.

3. This titration generally requires the addition of 8 mL of aqueous, saturated sodium carbonate solution.

4. The literature value for this ester is reported to be $[\alpha]_D$ -8.7° (EtOH, c 1.13).[2] The NMR spectra are as follows: ^{1}H NMR (300 MHz, CDCl$_3$) δ: 2.18-2.56 (m, 4 H), 3.78 (s, 3 H), 4.26 (dd, 1 H, J = 8.5, 5.1), 6.25 (s(br), 1 H); ^{13}C NMR (75 MHz, CDCl$_3$) δ: 25.3, 29.9, 53.1, 56.1, 173.3, 179.1.

5. Rhodium(II) acetate is obtained commercially (Degussa Corporation, Aldrich Chemical Company, Inc., AESAR/Johnson Matthey, or Johnson Matthey/Alfa Products) or prepared from rhodium(III) chloride hydrate (Johnson Matthey) by the standard literature procedure.[3]

6. Chlorobenzene was distilled under nitrogen from calcium hydride (-4 to +40 mesh) prior to use.

7. A micro Soxhlet extraction apparatus (Ace Glass, Inc.) consisting of the extractor, a 50-mL round-bottomed flask, and an Allihn condenser fitted with a 10 x 50 mm thimble is used.

8. Sodium carbonate is used to trap acetic acid liberated in the ligand exchange reaction. Sand is included to maintain the porosity of the solid. A layer of sand at the top of the carbonate mixture prevents sodium carbonate from entering the reaction flask.

9. When the reaction is followed by HPLC using a μ-Bondapak-CN column with 2% acetonitrile in methanol as the eluent, two bands are observed initially: a broad band eluting with the solvent front [chlorobenzene and excess methyl 2-pyrrolidone-5(R)-carboxylate] and a second band at 1.6 min when the flow rate is 1.5 mL/min [rhodium(II) acetate]. As the reaction progresses, the rhodium(II) acetate band diminishes and is replaced by several bands with longer retention volumes until one major band, in addition to that for chlorobenzene and ligand, is observed at about 4 min. Only minor impurities elute at intermediate times. A brown-black material, insoluble in all common solvents, is observed in some preparations. The origin of this material is unknown, but its presence decreases product yield by 25%.

10. Unreacted methyl 2-pyrrolidone-5(R)-carboxylate may be reisolated by distillation (70% recovery) without loss of optical purity.

11. This material is >95% pure by HPLC analysis.

12. Direct recrystallization is an alternative to chromatography, but decreased yields result because of the large amount of unreacted ligand present. In this procedure 1 mL of acetonitrile is added to the crude, blue glass-like solid. The resulting solution is filtered through glass wool and 5 mL of isopropyl alcohol is added. Overnight refrigeration gives a 38% yield of red crystals that are identical with those described in the procedure.

13. This procedure produced analytically pure crystals with the following physical properties: Anal. Calcd for $C_{31}H_{46}N_6O_{13}Rh_2$: C, 40.63; H, 5.06; N, 9.17. Found: C, 40.37, H, 5.11; N, 9.12; ^{1}H NMR (300 MHz, $CDCl_3$) δ: 1.21 (d, 6 H, J = 6.1), 1.35 (d, 1 H, J = 4.4), 1.8-2.4 (m, 12 H), 2.26 (s, 6 H), 2.55-2.70 (m, 4 H), 3.68 (s, 6 H), 3.70 (s, 6 H), 3.95 (dd, 2 H, J = 8.6, 2.1) 3.96-4.08 (m, 1 H), 4.32 (dd, 2 H, J = 8.8, 3.0); ^{13}C NMR (75 MHz, $CDCl_3$) δ: 2.9, 25.0, 25.2, 26.0, 31.2, 31.4, 51.6, 51.9, 64.0, 66.4, 66.6, 114.9, 175.5, 175.7, 188.3, 188.5. $[\alpha]_D^{24}$ +270.6° (CH_3CN, c 0.112).

14. Tetrahydrofuran was distilled from lithium aluminum hydride.

15. This reaction has an induction period. If too much diketene is added before condensation begins, an exotherm may result. Reaction onset is evident by a color change and a significant increase in temperature.

16. The NMR spectra are as follows: ^{1}H NMR (300 MHz, CDCl$_3$) δ: 1.72 (s, 3 H), 1.79 (s, 3 H), 2.27 (s, 3 H), 3.45 (s, 2 H), 4.64 (d, 2 H, J = 7.3), 5.32-5.40 (m, 1 H); enol form at 1.96 (s) and 4.99 (s). ^{13}C NMR (75 MHz, CDCl$_3$) δ: 18.5, 26.2, 30.5, 50.5, 62.6, 118.6, 140.2, 167.7, 201.2 (minor amounts of the enol tautomer may also be present).

17. Commercially available p-acetamidobenzenesulfonyl azide (Aldrich Chemical Company, Inc.) was preferred over p-dodecylbenzenesulfonyl azide (Danheiser, R. L.; Miller, R. F.; Brisbois, *Org. Synth.* **1995**, *73*, 134, Note 9; Ref. 4a), or methanesulfonyl azide[4] for reasons of safety, yield, and ease of manipulation.

18. After this time, no starting β-keto ester is observed upon analysis by TLC.

19. The principal reaction competing with deacylation is ester hydrolysis.

20. The spectral properties are as follows: ^{1}H NMR (300 MHz, CDCl$_3$) δ: 1.72 (s, 3 H), 1.77 (s, 3 H), 4.66 (d, 2 H, J = 7.3), 4.74 (s, 1 H), 5.32-5.39 (m, 1 H); ^{13}C NMR (75 MHz, CDCl$_3$) δ: 18.4, 26.2, 46.6, 62.2, 119.2, 139.7, 167.4; IR (thin film) cm^{-1}: 2109 (C=N$_2$), and 1694 (C=O).

21. Dichloromethane was distilled from phosphorus pentoxide and stored under nitrogen. All glassware was oven dried.

22. The NMR spectra are as follows: ^{1}H NMR (300 MHz, CDCl$_3$) δ: 1.17 (s, 3 H), 1.18 (s, 3 H), 1.93 (d, 1 H, J = 6.2), 2.04 (dd, 1 H, J = 6.2, 5.5), 4.13 (d, 1 H, J = 9.8), 4.35 (dd, 1 H, J = 9.8, 5.5); ^{13}C NMR (75 MHz, CDCl$_3$) δ: 14.2, 25.0, 29.9, 30.3, 66.4, 174.9. Enantiomerically pure (1R,5S)-(-)-6,6-dimethyl-3-oxabicyclo[3.1.0]hexan-2-one is reported to have $[\alpha]_D^{25}$ -89.9° (CHCl$_3$, c 1.4) (calculated).[5]

23. A 30 m x 0.32 mm ID Chiraldex B-PH (β-cyclodextrin) column was used under isothermal conditions at 120°C. Retention time of the (1R,5S)-enantiomer was 16.9 min, while the (1S,5R)-enantiomer eluted at 17.9 min. The submitters report that separation also occurred on a 30-m Chiraldex γ-cyclodextrin trifluoroacetate column operated at 140°C; retention times were 14.2 min (1R,5S-enantiomer) and 18.9 min (1S,5R-enantiomer).

24. The submitters report that the recovered catalyst can be reused in the same reaction; (1R,5S)-(-)-6,6-dimethyl-3-oxabicyclo[3.1.0]hexan-2-one was formed in 83% yield and 88% ee.

Waste Disposal Information

All toxic materials were disposed of in accordance with "Prudent Practices for Disposal of Chemicals from Laboratories"; National Academic Press; Washington, DC, 1983.

3. Discussion

This is the first detailed procedure for the synthesis of a chiral dirhodium(II) carboxamide catalyst and its application to intramolecular cyclopropanation. The preparation of the ligand, methyl 2-pyrrolidone-5(R)-carboxylate, is adapted from the procedure of Ackermann, Matthes, and Tamm.[2] The method for ligand displacement from dirhodium(II) tetraacetate is an extension of that reported for the synthesis of dirhodium(II) tetraacetamide.[6] The title compound, (1R,5S)-(-)-6,6-dimethyl-3-oxabicyclo[3.1.0]hexan-2-one, is a synthetic precursor to (1R,3S)-(+)-cis-chrysanthemic acid.[5]

21

Dirhodium(II) tetrakis[methyl 2-pyrrolidone-5(R)-carboxylate], $Rh_2(5R\text{-MEPY})_4$, and its enantiomer, $Rh_2(5S\text{-MEPY})_4$, which is prepared by the same procedure, are highly enantioselective catalysts for intramolecular cyclopropanation of allylic diazoacetates (65-≥94% ee) and homoallylic diazoacetates (71-90% ee),[7,8] intermolecular carbon-hydrogen insertion reactions of 2-alkoxyethyl diazoacetates (57-91% ee)[9] and N-alkyl-N-(tert-butyl)diazoacetamides (58-73% ee),[10] intermolecular cyclopropenation of alkynes with ethyl diazoacetate (54-69% ee) or menthyl diazoacetates (77-98% diastereomeric excess, de),[11] and intermolecular cyclopropanation of alkenes with menthyl diazoacetate (60-91% de for the cis isomer, 47-65% de for the trans isomer).[12] Their use in ≤1.0 mol % in dichloromethane solvent effects complete reaction of the diazo ester and provides the carbenoid product in 43-88% yield. The same general method used for the preparation of $Rh_2(5R\text{-MEPY})_4$ was employed for the synthesis of their isopropyl[7] and neopentyl[9] ester analogs.

1. Department of Chemistry, Trinity University, San Antonio, TX 78212.

2. Ackermann, J.; Matthes, M.; Tamm, C. *Helv. Chim. Acta* **1990**, *73*, 122.

3. Rempel, G. A.; Legzdins, P.; Smith, H.; Wilkinson, G. *Inorg. Synth.* **1972**, *13*, 90.

4. Boyer, J. H.; Mack, C. H.; Goebel, W.; Morgan, L. R., Jr. *J. Org. Chem.* **1958**, *23* 1051.

5. Mukaiyama, T.; Yamashita, H.; Asami, M. *Chem. Lett.* **1983**, 385.

6. Doyle, M. P.; Bagheri, V.; Wandless, T. J.; Harn, N. K.; Brinker, D. A.; Eagle, C. T.; Loh, K.-L. *J. Am. Chem. Soc.* **1990**, *112*, 1906.

7. Doyle, M. P.; Pieters, R. J.; Martin, S. F.; Austin, R. E.; Oalmann, C. J.; Müller, P. *J. Am. Chem. Soc.* **1991**, *113*, 1423.

8. Martin. S. F.; Oalmann, C. J.; Liras, S. *Tetrahedron Lett.* **1992**, *33*, 6727.

9. Doyle, M. P.; van Oeveren, A.; Westrum, L. J.; Protopopova, M. N.; Clayton, Jr., T. W. *J. Am. Chem. Soc.* **1991**, *113*, 8982.

10. Doyle, M. P.; Protopopova, M. N.; Winchester, W. R.; Daniel, K. L. *Tetrahedron Lett.* **1992**, *33*, 7819.

11. Protopopova, M. N.; Doyle, M. P.; Müller, P.; Ene, D. *J. Am. Chem. Soc.* **1992**, *114*, 2755.

12. Doyle, M. P.; Brandes, B. D.; Kazala, A. P.; Pieters, R. J.; Jarstfer, M. B.; Watkins, L. M.; Eagle, C. T. *Tetrahedron Lett.* **1990**, *31*, 6613.

Appendix
Chemical Abstracts Nomenclature (Collective Index Number); (Registry Number)

(1R,5S)-(-)-6,6-Dimethyl-3-oxabicyclo[3.1.0]hexan-2-one: 3-Oxabicyclo[3.1.0]hexan-2-one, 6,6-dimethyl-, (1R-cis)- (10); (71565-25-8)

Dirhodium(II) tetrakis[methyl 2-pyrrolidone-5(R)-carboxylate: $Rh_2(5R-MEPY)_4$: Rhodium, tetrakis[μ-(methyl 5-oxo-L-prolinato-N^1: O^5)]di-, (Rh-Rh) (12); (1324-35-65-5)

Thionyl chloride (8,9); (7719-09-7)

Methyl 2-pyrrolidone-5(R)-carboxylate: D-Proline, 5-oxo-, methyl ester (10); (64700-65-8)

(R)-(+)-2-Pyrrolidone-5-carboxylic acid: D-Proline, 5-oxo- (8,9); (4042-36-8)

Rhodium(II) acetate: Acetic acid, rhodium(2+) salt (8,9); (5503-41-3)

Acetonitrile: TOXIC: (8,9); (75-05-8)

Isopropyl alcohol: 2-Propanol (8,9); (67-63-0)

3-Methyl-2-buten-1-ol: 2-Buten-1-ol, 3-methyl- (8,9); (556-82-1)

Diketene: 2-Oxetanone, 4-methylene- (8,9); (674-82-8)

Triethylamine (8); Ethanamine, N,N-diethyl- (9); (121-44-8)

p-Acetamidobenzenesulfonyl azide: Sulfanilyl azide, N-acetyl- (8);

Benzenesulfonyl azide, 4-(acetylamino)- (9); (2158-14-7)

Rhodium(III) chloride hydrate: Rhodium chloride, hydrate (8,9); (20765-98-4)

ENANTIOSELECTIVE HYDROLYSIS OF cis-3,5-DIACETOXYCYCLOPENTENE: (1R,4S)-(+)-4-HYDROXY-2-CYCLOPENTENYL ACETATE

(4-Cyclopentene-1,3-diol, monoacetate, (1R-cis)-)

(±)-**1**   **2**   (+)-**1**

Submitted by Donald R. Deardorff, Colin Q. Windham, and Chris L. Craney.[1]

Checked by Renaud Beaudegnies and Leon Ghosez.

1. Procedure

A. *cis-3,5-Diacetoxycyclopentene.* A flame-dried, 100-mL, single-necked, round-bottomed flask is equipped with a serum cap, Teflon-coated magnetic stirring bar, and an 18-gauge needle attached to a dry source of argon vented through a mineral oil bubbler. The apparatus is flushed with argon and charged with 16.70 g (118 mmol) of (±)-cis-4-acetoxy-1-hydroxycyclopent-2-ene [(±)-**1**] (Note 1) and 20 mL of methylene chloride (Note 2). To this stirred solution is added 8.81 g (129 mmol) of imidazole in one portion (Note 3). When dissolution is complete, the reaction mixture is cooled to 0°C with the aid of an ice-water bath. Acetic anhydride (13.20 g, 12.20 mL, 129 mmol) (Note 4) is added dropwise over 5 min by means of an oven-dried syringe and needle. The cooling bath is removed and the reaction is allowed to stir at ambient temperature until judged complete by TLC analysis (Note 5). The contents of the flask are transferred to a 250-mL separatory funnel and diluted with 60 mL of reagent grade ethyl ether. The contents of the separatory funnel are washed with 60 mL of ice-cold 1 N hydrochloric acid which results in noticeable removal of all color from the organic phase. The acidic aqueous phase is back-extracted with two 50-mL portions of ether (Note 6). The combined organic phases are washed with a single 50-mL portion of saturated sodium bicarbonate solution followed by back-extraction of the aqueous phase with three 50-mL portions of ether (Note 6). The combined organic extracts are dried over anhydrous magnesium sulfate, filtered, and concentrated under rotary evaporation at 30 mm. The resulting viscous tan oil is distilled through a 10-cm Vigreux column to afford 20.9-21.1 g (96-98%) of cis-3,5-diacetoxycyclopentene (**2**) (Note 7) as a colorless liquid, bp 55-56°C at 0.1 mm.

B. *(1R,4S)-(+)-4-Hydroxy-2-cyclopentenyl acetate.* A rigorously clean 1-L Erlenmeyer flask, equipped with a 2-in. long Teflon-coated stirring bar (Note 8), is charged with 320 mL of a 1.45 M sodium dihydrogen phosphate buffer concentrate (Note 9). The solution is diluted to a final volume of 800 mL with the addition of 480 mL of glass-distilled water. To the gently stirred solution (Note 10) is added 78 mg of sodium azide (Note 11) followed by 18.6 mg (200 units/mg) of lyophilized electric eel acetyl cholinesterase (EEAC) (Note 12). The enzyme completely dissolves within 5 min after which time 16.01 g (87.0 mmol) of cis-3,5-diacetoxycyclopentene is added in one portion. A few extra mL of distilled water are used to facilitate a quantitative transfer of the diacetate. Parafilm is placed over the mouth of the flask and the two-phase system (Note 13) is allowed to stir at room temperature (Note 14). The progress of the reaction is conveniently monitored by periodic TLC analysis. The reaction is terminated when only a trace of diacetate remains and the corresponding diol begins to appear (Note 15). The aqueous phase is first extracted three times with 150-mL portions of ether followed by fifteen 150-mL extractions with a 1:1 mixture of ether/ethyl acetate (Note 16). The organic extracts are combined, dried over anhydrous magnesium sulfate, and concentrated by rotory evaporation at ~30 mm (Note 17). The

26

residue is passed through a 60-g plug of silica gel (Note 18) with approximately 750 mL of ether. Removal of the solvent by rotary evaporation followed by exposure to oil pump vacuum affords 11.32 g of nearly colorless material [mp 43-49°C (corrected)]. Vacuum distillation (Note 19) of the product through a short-path apparatus at 64°C/0.060 mm furnishes 10.64-10.76 g (86-87%, 96% e.e.) of colorless solid (+) hydroxy acetate **1** (mp 44.5-49.5°C (corrected), $[\alpha]_D^{23}$ +62.9° (chloroform, *c* 1.465)) (Note 20). A sample recrystallized from 1:1 pentane/ether (approximately 5 mL/g) provides high purity (≥99% e.e.), colorless needles of **1** (mp 50.7-51.3°C (corrected), $[\alpha]_D^{23}$ +73.8° (chloroform, *c* 1.25) [lit.[2] mp 49°-51°C, $[\alpha]_D^{23}$ +68.0° (chloroform, *c* 1.64)] (Note 21).

2. Notes

1. (±)-cis-4-Hydroxy-2-cyclopentenyl acetate [(±)-**1**] was prepared according to the method described in *Org. Synth.*[3] This material must be recrystallized (1:1 ether-pentane) before use to remove impurities (e.g., trans-monoacetate) that otherwise lead to reduced optical rotations in the final product [(+)-**1**]. The submitters have found that material having a melting point of 37-39°C is of satisfactory purity. The racemic monoacetate is hygroscopic so the proper precautions should be taken to prevent exposure to moisture. The NMR spectrum is as follows: ^{1}H NMR of (±) **1** (200 MHz, CDCl$_3$) δ: 1.64 (dt, 1 H, J = 14.7, 3.7), 2.02 (s, 3 H), 2.78 (overlapping dt, 1 H, J = 14.7, 7.5), 4.70 (overlapping s, 1 H and broad s), 5.45 (m, 1 H), 5.95 (m, 1H), 6.10 (m, 1H); ^{13}C NMR of (+) **1** (50 MHz, CDCl$_3$) δ: 21.2, 40.5, 74.9, 77.0, 132.6, 138.5, 170.8.

2. HPLC grade methylene chloride was purchased from Fisher Scientific Company and distilled from calcium hydride through a Vigreux column prior to use.

3. Imidazole (99% indicated purity) was purchased from Aldrich Chemical Company, Inc., and used without any further purification.

4. Acetic anhydride (98% indicated purity) was purchased from Aldrich Chemical Company, Inc., and distilled from phosphorus pentoxide through a Vigreux column prior to use.

5. Thin layer chromatographic (TLC) analysis was performed on Baker Si250F precoated plates with ethyl ether as the solvent. The approximate R_f values for the monoacetate and diacetate under these conditions are 0.44 and 0.72, respectively. The compounds were visualized using p-anisaldehyde stain.

6. This step is necessary to retrieve the slightly water-soluble product diester that migrates into the aqueous layer during the wash process. The meso-diester is easily detected in the water phase by TLC analysis (see Note 5).

7. cis-3,5-Diacetoxycyclopentene has the following spectral data: ^{1}H NMR of **2** (200 MHz, CDCl$_3$) δ: 1.74 (dt, 1 H, J = 15, 3.85), 2.05 (s, 6 H), 2.87 (overlapping dt, 1 H, J = 15, 7.6), 5.55 (m, 2 H), 6.1 (s, 2 H); ^{13}C NMR of **2** (50 MHz, CDCl$_3$) δ: 21.1, 37.1, 76.6, 134.6, 170.7.

8. The stirring bar should be free of any slivers of magnetic material or heavy metals that could inactivate the enzyme.

9. The buffer was prepared by dissolving 100 g of sodium dihydrogen phosphate monohydrate (Alfa Johnson Matthey) into 200 mL of glass-distilled water. The pH of the solution was adjusted to 6.9 with the addition of concentrated sodium hydroxide solution (MCB Reagent) and verified with a pH meter recently calibrated between 4-7 pH units using Fisher standards. The buffer solution was diluted to a final volume of 500 mL (1.45 M) and stored in a clean plastic bottle.

10. The stirring rate was adjusted so as to minimize the size of the vortex, which would introduce air into the solution and jeopardize the enzyme.

11. Sodium azide (Sigma Chemical Company) was used to protect the solution's components from microbial attack during the prolonged reaction period.

12. Lyophilized electric eel acetyl cholinesterase (C-3389, EC 3.1.1.7) was purchased from Sigma Chemical Company and stored below 0°C.

13. As the reaction nears completion, all the diacetate dissolves.

14. The success of this reaction may be temperature dependent. Shortly after starting this reaction on a particularly hot day, the building air conditioning failed. The temperature of the reaction flask quickly reached 28.7°C. The optical rotation of the distilled product was 4° off and the melting point was lower than expected. In all reactions prior to or since that incident the reaction temperatures never climbed above 23°C and the results have been uniformly superior.

15. Under these conditions, the reaction lasts 9-12 hr with a final pH of approximately 6.5. Nonspecific hydrolysis of 2 over this time accounts for less than 0.5% of the product as measured by control experiments. If cost is of little concern, the reaction time can be shortened dramatically by the addition of extra enzyme. cis-3,5-Dihydroxycyclopentene has an R_f value in ether of approximately 0.1.

16. HPLC grade ethyl acetate was purchased from J. T. Baker Inc. and distilled from phosphorus pentoxide prior to use. Although the product, (1R,4S)-(+)-4-hydroxy-2-cyclopentenyl acetate [(+)-1], is more soluble in ethyl acetate, emulsion concerns dictated the use of pure ether for the first three extractions and then a 1:1 ether/ethyl acetate mixture for all subsequent extractions. When emulsions did occur during the extraction process, the unemulsified portion of the aqueous phase was drained and a few mL of brine solution was added to the emulsified phase. Two or three quick shakes alleviated the problem.

17. The residual oil may crystallize during this process. If so, it will be necessary to dissolve the solid in about 10 mL of ether (with the aid of a hot air gun) prior to passing the material through the silica gel plug.

29

18. Baker Analyzed Reagent silica gel 40-140 mesh was used in a 4.5-cm diameter column. This step removes the small amounts of diol present. TLC analysis was helpful in determining when all the monoacetate had been stripped from the plug.

19. The receiving flask should be submerged in an ice-water bath since the product has a considerable vapor pressure under these conditions. It is wise to have a heat gun nearby in the event that the condensate begins to crystallize in the condenser.

20. The enantiomeric excess was unambiguously determined via gas chromatographic separation of diastereomers obtained from derivation of the chiral alcohol with Mosher's reagent,[4] (R)-(+)-α-methoxy-α-(trifluoromethyl)phenylacetic acid, (MTPA) 99% (Aldrich Chemical Company, Inc.). The checkers used the procedure of Noyori: Kitamura, M.; Tokunaga, M.; Ohkuma, T., Noyori, R. *Org. Synth.* **1992**, *71*, 1, for derivatization with Mosher's reagent. The MTPA esters derived from the (-)- and (+)-monoacetate enantiomers have retention times of 108 and 111.5 min, respectively, on a 25-m OV-101, crosslinked, methylsilicone capillary column operating at 150°C and flow rate of 22 cm/sec. Although less reliable than GC, the ratio of MTPA esters can be estimated via the use of high field (≥200 MHz) NMR in $CDCl_3$ solvent.[5]

21. One paper reports[5] an optical rotation of $[\alpha]_D^{22}$ +75.0° (chloroform, *c* 1.16) for crystals of (+)-**1** melting at only 47.5-48°C. We find these data to be inconsistent with our own experience and with the absolute values of the reported rotations for optically pure (-)-**1**. The polarimeter used in the submitter's analysis, a JACSO DIP-360, had been previously calibrated with NBS Standard Reference Sucrose 17d in accordance with NBS guidelines.

Waste Disposal Information

All toxic materials were disposed of in accordance with "Prudent Practices for Disposal of Chemicals from Laboratories"; National Academic Press; Washington, DC, 1983.

3. Discussion

In the 1980's there was a great increase in the development and use of enzymatic procedures by synthetic chemists.[6] Previously regarded more as scientific curiosities of limited scope than of practical utility, biological-chemical transformations are now used regularly by synthetic chemists. The ability to induce optical activity in molecules where none existed before is the most useful property of these chiral catalysts. Hydrolase enzymes are generally preferred over other kinds of enzymes for transformations of this nature because they are more easily handled and do not require cofactors for activity. In cases where enantiotopic differentiation between ester functions is desired, prochiral meso diesters are more efficient substrates than racemic esters. In the former case it is possible for all starting material to be converted into a single enantiomer, whereas in the latter example only enzymatic resolution is possible.

(1R,4S)-(+)-4-Hydroxy-2-cyclopentenyl acetate [(+)-1] is an important synthetic precursor. It provides optically active starting material via the versatile intermediate 4-oxo-2-cyclopentenyl acetate,[7] for important cyclopentanoids such as the prostaglandins[8] and carbocyclic nucleosides.[9] Because of the medicinal significance of these compounds more efficient routes, with better enantioselectivities have been devised to nonracemic 1. Enzymatic catalysis has become the dominant methodology for induction of this optical activity.

In addition to the title procedure,[10] other enzymatic preparations of nearly optically pure (1R,4S)-(+)-4-hydroxy-2-cyclopentenyl acetate [(+)-1][5,11] and its optical antipode (-)-1[12,13,14,15,16] are known. These enzyme-catalyzed procedures are derivatives of two basic strategies: (1) the enantioselective hydrolysis of the meso-diacetate 2,[5,10,11,12,13] or (2) the enantioselective transacetylation of the parent meso-diol.[14,15,16] Although (-)-1 has been successfully prepared by either route, the (+)-enantiomer is available only via the hydrolytic approach.

The other enzymatic routes[5,11] to (+)-1 follow roughly similar procedures. Both employ pH 7 buffer solutions of the hydrolytic enzyme porcine pancreas lipase (PPL) as opposed to electric eel acetyl cholinesterase (EEAC) used here. Although EEAC does afford better enantioselection than PPL (96% vs. 92% e.e.[11]), this apparent advantage may be viewed as somewhat tenuous since (+)-1 can be optically enriched via repeated recrystallizations. However, EEAC is clearly superior to PPL in terms of chemical yield. PPL mediated hydrolyses lead to yields of (+)-1 in the 50% range (plus recovered starting material), while the EEAC conversion rate is 90-95%. More importantly, the EEAC route eliminates the need to recycle starting material.

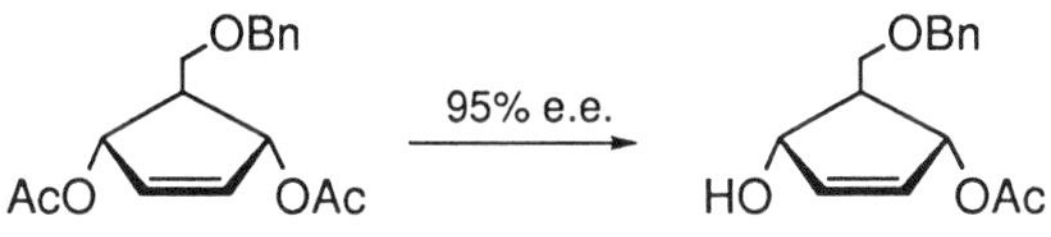

OBn ... 95% e.e. ... OBn

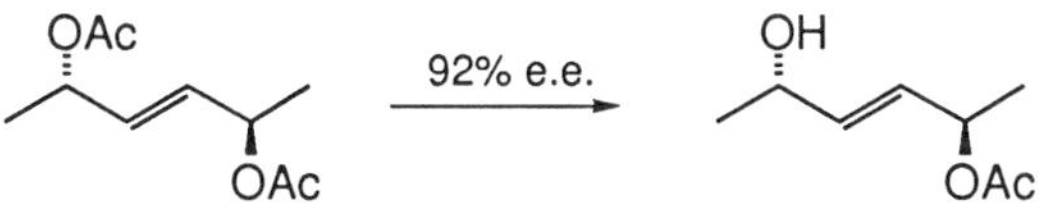

AcO ... OAc (7) → HO ... OAc (8)

OAc ... 92% e.e. ... OH

OAc (9) → OAc (10)

EEAC has been used successfully in other enantiotopic differentiations. Johnson, et al.[17] have reported that diester **3** can be readily transformed into hydroxy acetate **4** via this enzymatic process in 98% e.e. and an 80% chemical yield. Similarly, hydroxy acetate **6** was prepared from its parent diester **5** by Pearson, et al.[18] in 100% e.e., although 39% yield (50-55% recovered starting material). The enzyme also appears effective on 4-substituted cis-3,5-diacetoxycyclopentenes as Danishefsky[19] demonstrated with the conversion of **7** into **8** in 95% yield and 95% e.e. Finally, the successful enantioselective hydrolysis of **9** into **10** (77%, 92% e.e.) extends the range of useful EEAC substrates to acyclic cases.[20]

Immobilized EEAC (Sigma Chemical Company) has been found equally effective and allows economical, large-scale hydrolyses of **2**. The immobilized enzyme, which is covalently bonded to agarose beads, is easily recovered through centrifugation. Reuse does not compromise the enantioselectivity (unpublished results).

1. Department of Chemistry, Occidental College, Los Angeles, CA 90041. We gratefully acknowledge the National Science Foundation (CHE-8908212, CHE-8804037 & DMB-9005512) for financial support of this work. C. Q. W. wishes to thank the Ford Foundation for his summer undergraduate research stipend.

2. Busato, S.; Tinembart, O.; Zhang, Z. D.; Scheffold, R. *Tetrahedron* **1990**, *46*, 3155.

3. Deardorff, D. R.; Myles, D. C. *Org. Synth.* **1989**, *67*, 114.

4. Dale, J. A.; Dull, D. L.; Mosher, H. S. *J. Org. Chem.* **1969**, *34*, 2543.

5. Sugai, T.; Mori, K. *Synthesis* **1988**, 19.

6. For recent reviews see: (a) Chen, C.-S.; Sih, C. J. *Angew. Chem., Int. Ed. Engl.* **1989**, *28*, 695; (b) Ohno, M.; Otsuka, M. *Org. React.* **1989**, *37*, 1; (c) Jones, J. B.; *Tetrahedron* **1986**, *42*, 3351; (d) Whitesides, G. M.; Wong, C.-H. *Angew. Chem., Int. Ed. Engl.* **1985**, *24*, 617; (e) Suckling, C. J. In "Enzyme Chemistry, Impact and Applications;" Suckling, C. J., Ed., Chapman and Hall, New York, **1984**, pp. 78-118; (f) Whitesides, G. M.; Wong, C.-H. *Aldrichim. Acta* **1983**, *16*, 27.

7. Harre, M.; Raddatz, P.; Walenta, R.; Winterfeldt, E. *Angew. Chem., Int. Ed. Engl.* **1982**, *21*, 480.

8. Noyori, R.; Suzuki, M. *Angew. Chem., Int. Ed. Engl.* **1984**, *23*, 847.

9. Deardorff, D. R.; Shambayati, S.; Myles, D. C.; Heerding, D. *J. Org. Chem.* **1988**, *53*, 3614.

10. Deardorff, D. R.; Matthews, A. J.; McMeekin, D. S.; Craney, C. L. *Tetrahedron Lett.* **1986**, *27*, 1255.

11. Laumen, K.; Schneider, M. P. *J. Chem. Soc., Chem. Commun.* **1986**, 1298.

12. Wang, Y.-F.; Chen, C.-S.; Girdaukas, G.; Sih, C. J. *J. Am. Chem. Soc.* **1984**, *106*, 3695.

13. Laumen, K.; Schneider, M. *Tetrahedron Lett.* **1984**, *25*, 5875.

14. Theil, F.; Ballschuh, S.; Schick, H.; Haupt, M.; Häfner, B.; Schwarz, S. *Synthesis* **1988**, 540.

15. Jommi, G.; Orsini, F.; Sisti, M.; Verotta, L. *Gazz. Chim. Ital.* **1988**, *118*, 863.

16. Babiak, K. A.; Ng, J. S.; Dygos, J. H.; Weyker, C. L.; Wang, Y.-F.; Wong, C.-H. *J. Org. Chem.* **1990**, *55*, 3377.

17. Johnson, C. R.; Penning, T.D. *J. Am. Chem. Soc.* **1986**, *108*, 5655.

18. Pearson, A. J.; Bansal, H. S.; Lai, Y.-S. *J. Chem. Soc., Chem. Commun.* **1987**, 519.

19. Griffith, D. A.; Danishefsky, S. J. *J. Am. Chem. Soc.* **1991**, *113*, 5863.

20. Schink, H. E.; Bäckvall, J-E. *J. Org. Chem.* **1992**, *57*, 1588.

Appendix

Chemical Abstracts Nomenclature (Collective Index Number);

(Registry Number)

cis-3,5-Diacetoxycyclopentene: 4-Cyclopentene-1,3-diol, diacetate, cis- (9); (54664-61-8)

(1R,4S)-(+)-4-Hydroxy-2-cyclopentyl acetate: 4-Cyclopentene-1,3-diol, monoacetate, (1R-cis)- (9); (60410-16-4)

(±)-cis-4-Acetoxy-1-hydroxycyclopent-2-ene: 4-Cyclopentene-1,3-diol, monoacetate, cis- (9); (60410-18-6)

Imidazole (8); 1H-Imidazole (9); (288-32-4)

Acetic anhydride (8); Acetic acid anhydride (9); (108-24-7)

Sodium dihydrogen phosphate: Phosphoric acid, monosodium salt (8,9); (7558-80-7)

Sodium azide (8,9); (26628-22-8)

Acetyl cholinesterase (from electric eel): Esterase, acetyl choline (9); (9000-81-1)

(4R)-(+)-tert-BUTYLDIMETHYLSILOXY-2-CYCLOPENTEN-1-ONE

(2-Cyclopenten-1-one, 4-[[(1,1-dimethylethyl)dimethylsilyl]oxy]-, (R)-)

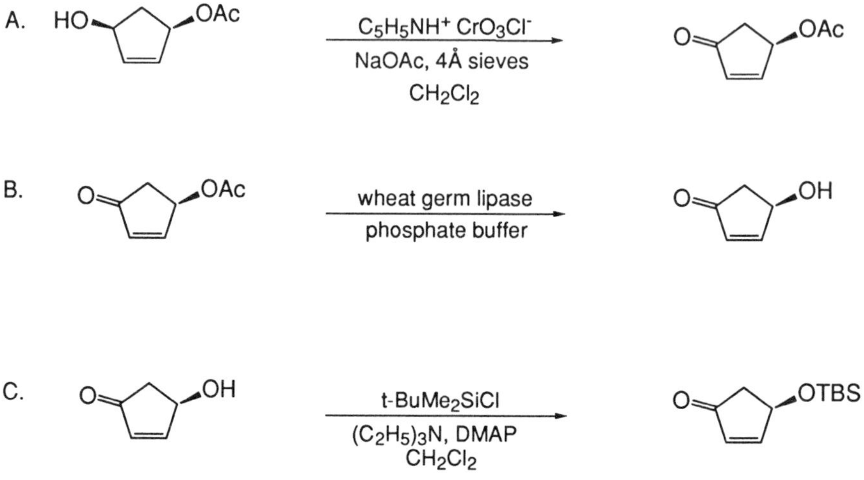

A. HO⋯△⋯OAc

$C_5H_5NH^+ CrO_3Cl^-$
NaOAc, 4Å sieves
CH_2Cl_2

O=△⋯OAc

B. O=△⋯OAc

wheat germ lipase
phosphate buffer

O=△⋯OH

C. O=△⋯OH

t-BuMe$_2$SiCl
$(C_2H_5)_3N$, DMAP
CH_2Cl_2

O=△⋯OTBS

Submitted by Leo A. Paquette, Martyn J. Earle, and Graham F. Smith.[1]

Checked by Thomas Kirrane and Albert I. Meyers.

1. Procedure

A. *(4R)-(+)-Acetoxy-2-cyclopenten-1-one.* A flame-dried, 2-L, three-necked, round-bottomed flask, equipped with a Teflon-coated magnetic stirring bar, is purged with nitrogen and charged with 1.0 L of dry dichloromethane (Note 1), 1.5 g of anhydrous sodium acetate (Note 2), 70 g of 4 Å molecular sieves (Note 3), and 23 g (162 mmol) of (1R,4S)-(+)-4-hydroxy-2-cyclopentenyl acetate (Note 4). Finely powdered pyridinium chlorochromate (50 g, 240 mmol) is added portionwise over a period of 5 min, and the mixture is stirred at room temperature for 3 hr (Note 5), then filtered through a pad of Florisil. The filtrate is concentrated on a rotary evaporator, and the residual dark oil is purified by flash chromatography on silica gel with elution by 20% ethyl acetate in petroleum ether (bp 35-60°C) (Note 6) to give 18.9 g (83%) of (4R)-(+)-acetoxy-2-cyclopenten-1-one as a colorless oil (Note 7).

B. *(4R)-(+)-Hydroxy-2-cyclopenten-1-one.* A 2-L, one-necked, round-bottomed flask, fitted with a Teflon-coated magnetic stirring bar, is charged with 1.5 L of 0.05 M phosphate buffer (Note 8) and 18.8 g (134 mmol) of (4R)-(+)-acetoxy-2-cyclopenten-1-one. A 4.0-g lot of wheat germ lipase (Note 9) is added slowly with rapid stirring. Once the enzyme is dispersed, the speed of the stirrer is reduced and the flask is sealed with a glass stopper prior to being stirred at room temperature for 7 days (Note 10). The stopper is removed and the contents of the flask are transferred to the body of a lighter than water continuous extraction apparatus whose pot is charged with 1 L of ethyl acetate. After extraction for 3 days, the ethyl acetate solution is concentrated on a rotary evaporator, and the residue is subjected to flash chromatography on silica gel using 30% ethyl acetate in petroleum ether (bp 35-60°C) as eluent (Note 11). After recovery of 4.60 g (25%) of the less polar, unreacted acetoxy ketone (Note 12), 7.80 g (60%) of the colorless, oily (R)-(+)-hydroxy ketone is obtained (Note 13).

C. *(4R)-(+)-tert-Butyldimethylsiloxy-2-cyclopenten-1-one.* A 500-mL, three-necked, round-bottomed flask equipped with a Teflon-coated magnetic stirring bar, pressure-equalizing addition funnel, and nitrogen inlet, is flame-dried under a stream of dry nitrogen. The apparatus is charged with 7.7 g (78 mmol) of (4R)-(+)-hydroxy-2-cyclopenten-1-one, 0.96 g (10 mol %) of 4-dimethylaminopyridine (Note 14), 20 g (200 mmol) of triethylamine (Note 2), and 150 mL of dry dichloromethane (Note 1). The dropping funnel is charged with a solution of 14.2 g (94 mmol) of tert-butyldimethylsilyl chloride (Note 14) in 50 mL of dry dichloromethane. Magnetic stirring is initiated, the reaction mixture is cooled in an ice-water bath, and the silyl chloride solution is added dropwise during 10 min. The ice-water bath is removed, and the mixture is stirred at room temperature for 3 hr; then 100 mL of deionized water is added with stirring. After the organic layer is separated, the aqueous phase is extracted three times with 50 mL of dichloromethane and the combined organic solutions are dried over anhydrous magnesium sulfate, filtered, and concentrated on a rotary evaporator. The residue is filtered through a short column of silica gel with elution by 5% ethyl acetate in petroleum ether (bp 35-60°C). After concentration of the fractions containing the product, the resulting oil is distilled in a short path distillation apparatus (bp 60°C at 0.1 mm) to give 13.2 g of a colorless oil. Crystallization of this material from pentane with cooling by an acetone-dry ice bath gives 10.6 g (64%) of colorless needles, mp 27-28°C, $[\alpha]_D^{20}$ +65.3° (CH$_3$OH, c 0.4) (Notes 15 and 16).

2. Notes

1. Dichloromethane is distilled from calcium hydride prior to use.

2. Sodium acetate and triethylamine were purchased from the J. T. Baker Chemical Company. Sodium acetate is used without further purification; triethylamine is distilled from calcium hydride prior to use.

3. The 4 Å molecular sieves were purchased from the Aldrich Chemical Company, Inc., and activated by drying in a vacuum oven at 150°C for 24 hr prior to storing in an oven at 140°C. Both powdered and pelletized forms of the sieve were used without affecting the yield of product.

4. High purity (≥99% ee) (1R,4S)-4-hydroxy-2-cyclopentenyl acetate exhibiting $[\alpha]_D^{23}$ values of +71.1° to +71.3° in CHCl$_3$ can be obtained by enzymatic hydrolysis of the racemic diacetate (*Org. Synth.* **1995**, *73*, 25)[2,3,4] with electric eel acetyl cholinesterase[4] or with A.K. lipase (Amano International Enzyme Company).[5]

5. The progress of this reaction is conveniently monitored by TLC. The silica gel plates are eluted with 50% ethyl acetate in petroleum ether (bp 35-60°C). Under these conditions, the alcohol exhibits an R$_f$ of 0.45 and the acetate an R$_f$ of 0.80.

38

6. The column consists of 300 g of silica gel; 100-mL sized fractions are collected. Since release of the product from the chromium salts occurs slowly, fractions 5-20 are found to contain the acetoxy ketone and are combined.

7. The spectral data for (4R)-(+)-acetoxy-2-cyclopenten-1-one are as follows: ^{1}H NMR (300 MHz, CDCl$_3$) δ: 2.06 (s, 3 H), 2.29 (dd, 1 H, J = 2.2, 18.7), 2.78 (dd, 1 H, J = 6.4, 18.7), 5.81 (m, 1 H), 6.29 (dd, 1 H, J = 1.3, 5.7), 7.53 (dd, 1 H, J = 2.4, 5.7); ^{13}C NMR (75 MHz, CDCl$_3$) δ: 20.7, 40.9, 71.9, 136.9, 158.8, 170.3, 204.7; IR (CHCl$_3$) cm^{-1}: 2950, 1740, 1720, 1600, 1375, 1355, 1190, 795; $[\alpha]_D^{20}$ +96.1° (CH$_3$OH, c 0.17) [lit.[6] $[\alpha]_D^{22}$ +97° (CH$_3$OH, c 0.1].

8. The buffer is prepared by dissolving 27.2 g of potassium dihydrogen phosphate in 4.0 L of deionized water and titrating with 1 M sodium hydroxide until pH 5.0 is reached.

9. Wheat germ lipase was purchased from the Sigma Chemical Company.

10. The reaction rate decreased significantly toward the end of the hydrolysis. The level of conversion can be improved either by increasing the amount of enzyme or by lengthening the reaction time. The course of the reaction can be monitored by TLC on silica gel (elution with 50% ethyl acetate in petroleum ether). The product has an R$_f$ of 0.20 and the starting material an R$_f$ of 0.40.

11. The column consists of 200 g of silica gel; 100-mL sized fractions are collected. The starting material was recovered in fractions 2-5 while the product was recovered from fractions 7-20.

12. This material can be recycled in future reactions.

13. The spectral data for (4R)-(+)-hydroxy-2-cyclopenten-1-one are as follows: ^{1}H NMR (300 MHz, CDCl$_3$) δ: 2.27 (dd, 1 H, J = 3.0, 18.0), 2.78 (dd, 1 H, J = 6.0, 18.0), 3.27 (br s, 1 H), 4.88-5.22 (m, 1 H), 6.25 (d, 1 H, J = 6.0), 7.60 (dd, 1 H, J = 2.0, 6.0); ^{13}C NMR (75 MHz, CDCl$_3$) δ: 44.1, 70.1, 134.7, 164.0, 207.3; $[\alpha]_D^{20}$ +78.1° (CH$_3$OH, c 2.03).

14. 4-Dimethylaminopyridine and tert-butyldimethylsilyl chloride were purchased from the Aldrich Chemical Company, Inc., and used without purification.

15. The spectral data for (4R)-(+)-tert-butyldimethylsiloxy-2-cyclopenten-1-one are as follows: ^{1}H NMR (300 MHz, CDCl$_3$) δ: 0.12 (s, 3 H), 0.13 (s, 3 H), 0.91 (s, 9 H), 2.23 (dd, 1 H, J = 2.3, 18.2), 2.69 (dd, 1 H, J = 6.0, 18.2), 4.98 (m, 1 H), 6.17 (dd, 1 H, J = 1.2, 5.7), 7.44 (dd, 1 H, J = 2.3, 5.7); ^{13}C NMR (75 MHz, CDCl$_3$) δ: -4.76, -4.74, 18.0, 25.7, 44.9, 70.9, 134.4, 163.7, 206.3.

16. The reported optical rotations are $[\alpha]_D^{21}$ +66.6° (CH$_3$OH, c 1.0),[7] $[\alpha]_D^{21}$ +67.0° (CH$_3$OH, c 0.12),[6] and $[\alpha]_D^{21}$ +67.3° (CH$_3$OH, c 0.82).[8]

Waste Disposal Information

All toxic materials were disposed of in accordance with "Prudent Practices for Disposal of Chemicals from Laboratories"; National Academic Press; Washington, DC, 1983.

3. Discussion

As a direct consequence of the quest for optically active prostaglandins,[9] derivatives of (R)-4-hydroxy-2-cyclopenten-1-one have come to be regarded as important chiral building blocks. Initial efforts to obtain these compounds in enantiomerically pure form involved the chemical modification of D-tartaric acid,[10] degradation of the fungal metabolite terrein,[11] ring contraction of 2,4,6-trichlorophenol with resolution,[9,12] chromatography of diastereomeric[13] or racemic hydroxy-protected derivatives,[14] and a multi-step conversion from glucose.[15] Subsequent discoveries that excellent kinetic resolution can be achieved either by asymmetric BINAL-H reduction of 4-cyclopentene-1,3-dione[16] or by enzymatic hydrolysis of the acetate[17] proved to be major advances.

More recently, the desymmetrization of cis-3,5-diacetoxycyclopent-1-ene by enantioselective monohydrolysis in the presence of various enzymes has been intensively investigated.[18-23] This approach is readily adaptable to a laboratory setting, is inexpensive, and is capable of straightforwardly delivering multigram quantities of high quality (>99% ee) product.

The reaction sequence shown here, which has been adapted from earlier literature reports, permits the convenient acquisition of (4R)-(+)-tert-butyldimethylsiloxy-2-cyclopenten-1-one, perhaps the most useful of the possible derivatives for further synthetic elaboration.[24] The companion synthesis of the useful (4S) enantiomer from the same starting material is described in the next procedure.[25]

1. Evans Chemical Laboratories, The Ohio State University, Columbus, OH 43210.

2. Korach, M.; Nielson, D. R.; Rideout, W. H. *Org. Synth., Coll. Vol. V* **1973**, 414.

3. Deardorff, D. R.; Myles, D. C. *Org. Synth., Coll. Vol. VIII* **1993**, 13.

4. Deardorff, D. R.; Windham, C. Q.; Craney, C. L. *Org. Synth.* **1995**, *73*, 25.

5. Smith, G. F.; Earle, M. J., unpublished results.

6. Gill, M.; Rickards, R. W. *Tetrahedron Lett.* **1979**, 1539.

7. Kitamura, M.; Kasahara, I.; Manabe, K.; Noyori, R.; Takaya, H. *J. Org. Chem.* **1988**, *53*, 708.

8. Kitano, Y.; Okamoto, S.; Sato, F. *Chem. Lett.* **1989**, 2163.

9. Noyori, R.; Suzuki, M. *Angew. Chem., Int. Ed. Engl.* **1984**, *23*, 847.

10. Ogura, K.; Yamashita, M.; Tsuchihashi, G.-i. *Tetrahedron Lett.* **1976**, 759.

11. Mitscher, L. A.; Clark, G. W., III; Hudson, P. B. *Tetrahedron Lett.* **1978**, 2553.

12. Gill, M.; Rickards, R. W. *J. Chem. Soc., Chem. Commun.* **1979**, 121.

13. Suzuki, M.; Kawagishi, T.; Suzuki, T.; Noyori, R. *Tetrahedron Lett.* **1982**, *23*, 4057.

14. Okamoto, Y.; Aburatani, R.; Kawashima, M.; Hatada, K.; Okamura, N. *Chem. Lett.* **1986**, 1767.

15. Torii, S.; Inokuchi, T.; Oi, R.; Kondo, K.; Kobayashi, T. *J. Org. Chem.* **1986**, *51*, 254.

16. Noyori, R.; Tomino, I.; Yamada, M.; Nishizawa, M. *J. Am. Chem. Soc.* **1984**, *106*, 6717.

17. Dohgane, I.; Yamachika, H.; Minai, M. *Yuki Gosei Kagaku* **1983**, *41*, 896.

18. Wang, Y.-F.; Chen, C.-S.; Girdaukas, G.; Sih, C. J. *J. Am. Chem. Soc.* **1984**, *106*, 3695.

19. Laumen, K.; Schneider, M. *Tetrahedron Lett.* **1984**, *25*, 5875.

20. Deardorff, D. R.; Matthews, A. J.; McMeekin, D. S.; Craney, C. L. *Tetrahedron Lett.* **1986**, *27*, 1255.

21. Sugai, T.; Mori, K. *Synthesis* **1988**, 19.

22. Babiak, K. A.; Ng, J. S.; Dygos, J. H.; Weyker, C. L.; Wang, Y.-F.; Wong, C.-H. *J. Org. Chem.* **1990**, *55*, 3377.

23. Johnson, C. R.; Bis, S. J. *Tetrahedron Lett.* **1992**, *33*, 7287 and early references cited therein.

24. For a review detailing the chemistry of the acetate, see Harre, M.; Raddatz, P.; Walenta, R.; Winterfeldt, E. *Angew. Chem., Int. Ed. Engl.* **1982**, *21*, 480.

25. Paquette, L. A.; Heidelbaugh, T. M. *Org. Synth.* **1995**, *73*, 44.

Appendix

Chemical Abstracts Nomenclature (Collective Index Number); (Registry Number)

(4R)-(+)-tert-Butyldimethylsiloxy-2-cyclopenten-1-one: 2-Cyclopenten-1-one, 4-[[(1,1-dimethylethyl)dimethylsilyl]oxy]-, (R)- (9); (61305-35-9)

(4R)-(+)-Acetoxy-2-cyclopenten-1-one: 2-Cyclopenten-1-one, 4-(acetyloxy)-, (R)- (9); (59995-48-1)

(1R,4S)-(+)-4-Hydroxy-2-cyclopentyl acetate: 4-Cyclopentene-1,3-diol, monoacetate, (1R-cis)- (9); (60410-16-4)

Pyridinium chlorochromate: CANCER SUSPECT AGENT: Pyridine, chlorotrioxochromate (1-) (9); (26299-14-9)

(4R)-(+)-Hydroxy-2-cyclopenten-1-one: 2-Cyclopenten-1-one, 4-hydroxy-, (R)- (9); (59995-47-0)

Wheat germ lipase: Lipase, triacylglycerol; (9001-62-1)

4-Dimethylaminopyridine: HIGHLY TOXIC: Pyridine, 4-(dimethylamino)- (8);
4-Pyridinamine, N,N-dimethyl- (9); (1122-58-3)

Triethylamine: Ethanamine, N,N-diethyl- (8,9); (121-44-8)

tert-Butyldimethylsilyl chloride: CORROSIVE, TOXIC: Silane, chloro(1,1-dimethylethyl)-dimethyl- (9); (18162-48-6)

(4S)-(-)-tert-BUTYLDIMETHYLSILOXY-2-CYCLOPENTEN-1-ONE

((4S)-(-)-tert-Butyldimethylsiloxy-2-cyclopenten-1-one)

A. HO⟋△⟍OAc → t-BuMe$_2$SiCl / (C$_2$H$_5$)$_3$N, DMAP / CH$_2$Cl$_2$ → TBSO⟋△⟍OAc

B. TBSO⟋△⟍OAc → 1. NaOCH$_3$, CH$_3$OH / 2. MnO$_2$, CH$_2$Cl$_2$ → TBSO⟋△=O

Submitted by Leo A. Paquette and Todd M. Heidelbaugh.[1]

Checked by Thomas Kirrane and Albert I. Meyers.

1. Procedure

A. (1R,4S)-(-)-4-tert-Butyldimethylsiloxy-2-cyclopentenyl acetate. A dry, 500-mL, three-necked, round-bottomed flask, equipped with a Teflon-coated magnetic stirring bar, rubber septum, and nitrogen inlet, is purged with nitrogen and charged with 7.67 g (54 mmol) of (1R,4S)-(+)-4-hydroxy-2-cyclopentenyl acetate (Note 1), 660 mg (5.4 mmol) of 4-dimethylaminopyridine (Note 2), 17 mL (122 mmol) of triethylamine (Note 3), and 175 mL of dichloromethane (Note 3). The reaction mixture is cooled to 0°C in an ice-water bath, and tert-butyldimethylsilyl chloride (10.24 g, 68 mmol) (Note 2) is introduced in one portion. The ice-water bath is removed and the mixture is allowed to warm to room temperature and stir for 3 hr. At this point, more silyl chloride is added if necessary (Note 4). After 5 hr, 200 mL of water is added, the mixture is transferred to a separatory funnel and the organic phase separated. The aqueous phase is extracted with three 100-mL portions of dichloromethane. The combined

44

organic layers are washed with 100 mL of saturated sodium bicarbonate solution and 100 mL of brine prior to drying over anhydrous magnesium sulfate. After filtration and solvent removal with a rotary evaporator, the residual solids are removed by filtration (Note 5), and the resulting yellow oil is purified by bulb-to-bulb distillation at 0.4-0.6 mm (pot temperature 80-100°C) to give 10.67 - 11.08 g (77 - 80%) of (1R,4S)-(-)-4-tert-butyldimethylsiloxy-2-cyclopentenyl acetate as a colorless liquid, $[\alpha]_D^{20}$ -1.32° (CH$_3$OH, c 1.52) (Note 6).

B. *(4S)-(-)-tert-Butyldimethylsiloxy-2-cyclopenten-1-one.* A dry, 500-mL, one-necked, round-bottomed flask, equipped with a Teflon-coated magnetic stirring bar, is purged with nitrogen and charged with 11.7 g (45.6 mmol) of (1R,4S)-(-)-tert-butyldimethylsiloxy-2-cyclopentenyl acetate and 250 mL of anhydrous methanol (Note 7) to which 4.94 g (91.5 mmol) of powdered sodium methoxide (Note 8) is added. The reaction mixture is stirred for 15 min at ambient temperature, freed of most of the methanol using a rotary evaporator, and taken up in 400 mL of dichloromethane. The solution is washed with three 200-mL portions of water, dried over anhydrous magnesium sulfate, filtered, and concentrated using a rotary evaporator, giving 11.2 g of crude allylic alcohol which is carried into the next reaction without further purification.

A 500-mL, round-bottomed flask, equipped with a Teflon-coated magnetic stirring bar, is charged with the 11.2 g of crude allylic alcohol obtained above and 300 mL of dichloromethane, and the resulting vigorously stirred solution is treated with 33 g of active manganese dioxide (380 mmol) (Note 9). Additional 2-5 g lots of the oxidant are added every 2-3 hr until the reaction is complete (Note 10). The reaction mixture is vacuum-filtered through a pad of diatomaceous earth, and the pad is washed with 200 mL of dichloromethane. The resulting clear filtrate is concentrated carefully using a rotary evaporator, and the residual oil is purified by bulb-to-bulb distillation at 0.3 mm (pot temperature 100°C) affording 8.43 - 8.71 g (87 - 90%) of

45

enone as a pale yellow oil that solidifies when cooled below 15°C. Crystallization of the crude product from pentane at -70°C gives (4S)-(-)-tert-butyldimethylsiloxy-2-cyclopenten-1-one as colorless needles having mp 32-33°C, $[\alpha]_D^{23}$ -65.1° (CH$_3$OH, c 0.94) (Note 11).

2. Notes

1. High purity (≥99% ee) (1R,4S)-4-hydroxy-2-cyclopentenyl acetate exhibiting $[\alpha]_D^{20}$ values of +71.1 to +71.3° in CHCl$_3$ can be obtained by enzymatic hydrolysis of the racemic diacetate either with electric eel cholinesterase[2] or with A.K. lipase (Amano International Enzyme Company).[3] The checkers employed the EEAC procedure.[2]

2. 4-Dimethylaminopyridine and tert-butyldimethylsilyl chloride were purchased from the Aldrich Chemical Company, Inc. and used without further purification.

3. Triethylamine and dichloromethane were distilled from calcium hydride before use.

4. The progress of the reaction is easily monitored by TLC analysis. Silyl chloride is added until the starting hydroxy acetate is no longer detected.

5. Filtration is performed only to prevent bumping during the ensuing distillation.

6. The spectral data are as follows: ^{1}H NMR (300 MHz, CDCl$_3$) δ: 0.09 (s, 6 H), 0.90 (s, 9 H), 1.57-1.65 (m, 1 H), 2.04 (s, 3 H), 2.75-2.85 (dt, 1 H, J = 7.3, 3.8), 4.69-4.73 (m, 1 H), 5.44-5.48 (m, 1 H), 5.87-5.98 (m, 2 H); ^{13}C NMR (75 MHz, CDCl$_3$) δ: -4.7, -4.6, 18.2, 21.1, 25.9, 41.2, 74.9, 77.4, 131.2, 138.9, 170.9; IR (neat) cm^{-1}: 2940, 2870, 1740, 1610, 1375, 1250.

7. The methanol was refluxed over magnesium turnings and a crystal of iodine under nitrogen for 3 hr prior to use.

8. Sodium methoxide was freshly prepared by adding sodium metal to methanol (*Caution: hydrogen evolution*), evaporation of the solvent, and vacuum drying of the white solid.

9. Manganese dioxide was prepared as described.[6]

10. The time required to achieve complete reaction varies from 20-48 hr depending on the activity of the manganese dioxide. The progress of the oxidation is easily monitored by TLC analysis on silica gel.

11. The spectral data are identical to those reported for the (4R) enantiomer in the accompanying procedure.[7]

Waste Disposal Information

All toxic materials were disposed of in accordance with "Prudent Practices for Disposal of Chemicals from Laboratories"; National Academic Press; Washington, DC, 1983.

3. Discussion

Although (1R,4S)-(+)- and (1S,4R)-(-)-4-hydroxy-2-cyclopentenyl acetate are both available by enzyme-promoted enantioselective hydrolysis,[8,9] different enzymes are, of course, required to achieve this stereochemical divergence. Economy would be realized if one of these enantiomeric products could serve as the starting point for the preparation of both antipodal forms of structurally more advanced intermediates. The importance of (4R)-(+)-[10,11] and (4S)-(-)-tert-butyldimethylsiloxy-2-cyclopenten-1-one[12] to prostaglandin synthesis is well established. The latent potential of these highly functionalized building blocks for the enantiospecific synthesis of other natural

47

products is beginning to emerge.[13,14] Use of the present procedure makes possible the direct, efficient acquisition of the 4S enantiomer from the same hydroxy acetate that serves as a convenient progenitor to the 4R isomer.[7] The synthetic route is closely similar to that outlined earlier by Danishefsky, Cabal, and Chow.[13]

1. Evans Chemical Laboratories, The Ohio State University, Columbus, OH 43210.

2. Deardorff, D. R.; Windham, C. Q.; Craney, C. L. *Org. Synth.,* **1995**, *73*, 25.

3. Smith, G. F.; Earle, M. J., unpublished results.

4. Korach, M.; Nielson, D. R.; Rideout, W. H. *Org. Synth., Coll. Vol. V* **1973**, 414.

5. Deardorff, D. R.; Myles, D. C. *Org. Synth., Coll. Vol. VIII* **1993**, 13.

6. Attenburrow, J.; Cameron, A. F. B.; Chapman, J. H.; Evans, R. M.; Hems, B. A.; Jansen, A. B. A.; Walker, T. *J. Chem. Soc.* **1952**, 1094.

7. Paquette, L. A.; Earle, M. J.; Smith, G. F. *Org. Synth.* **1995**, *73*, 36.

8. Schneider, M.; Laumen, K. *Ger. Offen.* DE 3 620 646 (1987); *Chem. Abstr.* **1988**, *109*, 21633t.

9. Babiak, K. A.; Ng, J. S.; Dygos, J. H.; Weyker, C. L.; Wang, Y.-F.; Wong, C.-H. *J. Org. Chem.* **1990**, *55*, 3377 and relevant references cited therein.

10. Johnson, C. R.; Bis, S. J. *Tetrahedron Lett.* **1992**, *33*, 7287 and relevant references cited therein.

11. Kitamura, M.; Kasahara, I.; Manabe, K.; Noyori, R.; Takaya, H. *J. Org. Chem.* **1988**, *53*, 708.

12. Kitano, Y.; Okamoto, S.; Sato, F. *Chem. Lett.* **1989**, 2163.

13. Danishefsky, S. J.; Cabal, M. P.; Chow, K. *J. Am. Chem. Soc.* **1989**, *111*, 3456.

14. Paquette, L. A.; Ni, Z.; Smith, G. F.; Earle, M. J. unpublished work.

15. Paquette, L. A.; Heidelbaugh, T. M. unpublished results.

Appendix

Chemical Abstracts Nomenclature (Collective Index Number); (Registry Number)

(4S)-(-)-tert-Butyldimethylsiloxy-2-cyclopenten-1-one: 2-Cyclopenten-1-one, 4-[[(1,1-dimethylethyl)dimethylsilyl]oxy]-, (S)- (9); (61305-36-0)

(1R,4S)-(-)-4-tert-Butyldimethylsiloxy-2-cyclopentyl acetate: 2-Cyclopenten-1-ol, 4-[[(1,1-dimethylethyl)dimethylsilyl]oxy]-, acetate, (1R-cis)- (12); (115074-47-0)

(1R,4S)-(+)-4-Hydroxy-2-cyclopentyl acetate: 4-Cyclopentene-1,3-diol, monoacetate, (1R-cis)- (9); (60410-16-4)

4-Dimethylaminopyridine: HIGHLY TOXIC: Pyridine, 4-(dimethylamino)- (8); 4-Pyridinamine, N,N-dimethyl- (9); (1122-58-3)

Triethylamine (8); Ethanamine, N,N-diethyl- (9); (121-44-8)

tert-Butyldimethylsilyl chloride: CORROSIVE, TOXIC: Silane, chloro(1,1-dimethyl-ethyl)dimethyl- (9); (18162-48-6)

STEREOSELECTIVE ALKENE SYNTHESIS via 1-CHLORO-1-
[(DIMETHYL)PHENYLSILYL]ALKANES and
α-(DIMETHYL)PHENYLSILYL KETONES: 6-METHYL-6-DODECENE

A. PhMe₂SiCl → 1) Li, THF; hexanal / 2) CCl₄, PPh₃, THF → product with Cl and SiMe₂Ph

B. Cl-SiMe₂Ph compound → Mg; CuBr·Me₂S; C₅H₁₁COCl → [PhMe₂Si ketone intermediate]

1) MeLi →
  2) p-TsOH → alkene product (trisubstituted)
  2) KH → alkene product (trisubstituted)

Submitted by Anthony G. M. Barrett, John A. Flygare, Jason M. Hill and
Eli M. Wallace.[1]

Checked by Shuyong Chen and Amos B. Smith, III.

1. Procedure

A. 1-Chloro-1-[(dimethyl)phenylsilyl]hexane. An oven-dried, 500-mL, round-
bottomed flask, equipped with a magnetic stirring bar and a rubber septum, is purged
with argon via an inlet hose equipped with a needle and an outlet hose equipped with

a needle leading to an oil bubbler. The flask is charged with 250 mL of dry tetrahydrofuran (THF) (Note 1) and 7.37 g of lithium wire (1.06 mol) cut into small pieces (Note 2). After cooling the reaction mixture to 0°C in a CryoCool bath (Note 3), 30 g (29.1 mL, 176 mmol) of phenyldimethylsilyl chloride is added via syringe and the reaction mixture is stirred at 0°C for 16 hr (Note 4).[2] Within 0.5 hr, the reaction mixture turns from clear and colorless to dark red. An oven-dried, 1-L, round-bottomed flask, equipped with a magnetic stirring bar and a rubber septum, is purged with argon via an inlet needle and an outlet needle to an oil bubbler, charged with 16.78 g (20.0 mL, 167.5 mmol) of hexanal and 400 mL of dry THF (Note 5), and cooled to -78°C with a dry ice-acetone bath. The solution of phenyldimethylsilyllithium in THF prepared above is then added via cannula. After addition, the red reaction mixture is warmed to 0°C and stirred for 1 hr. A saturated aqueous ammonium chloride solution (250 mL) is then added in one portion and the resulting mixture is poured into a 2-L separatory funnel containing 250 mL of diethyl ether. The organic layer is washed with saturated aqueous ammonium chloride (3 x 250-mL) and brine (250 mL), dried over $MgSO_4$, and concentrated under reduced pressure. Purification via flash column chromatography (silica gel 230-400 mesh, 450 g of oil, loaded with hexanes, eluant 10:1 hexanes:diethyl ether) yields a total of 32.6 g (82%) 1-[(dimethyl)phenylsilyl]-1-hexanol as a colorless oil (a second column with 250 g of silica gel may be required for rechromatography of tailing fractions).

An oven-dried, 1-L, round-bottomed flask, equipped with a magnetic stirring bar, is charged with 500 mL of dry THF and 1-[(dimethyl)phenylsilyl]-1-hexanol (32.6 g, 138 mmol). Carbon tetrachloride (53.20 g, 33.4 mL, 345.9 mmol) and triphenylphosphine (54.30 g, 207 mmol) are added and the flask is equipped with a condenser and a rubber septum and purged with argon via inlet needle and outlet needle to an oil bubbler.[3] The reaction mixture is heated to reflux under argon for 12 hr. After the reaction mixture is allowed to cool to room temperature, the volatiles are removed under reduced pressure, and the residue is triturated with hexanes (3 x 300-mL). Concentration of the extracts under reduced pressure and purification via flash column chromatography (silica gel 230-400 mesh, 450 g, eluant hexanes) yields 30.4-31.7 g of 1-chloro-1-[(dimethyl)phenylsilyl]hexane (71-74% from hexanal) as a colorless oil (Note 6)

B. *6-Methyl-6-dodecene.* An oven-dried, 1-L, three-necked, round-bottomed flask is equipped with a condenser, a 125-mL, pressure-equalizing addition funnel, a magnetic stirring bar, and a rubber septum. The flask is purged with argon and charged with 96.23 g (373 mmol) of magnesium bromide etherate (MgBr$_2$·Et$_2$O) (Note 2) and 250 mL of dry THF. A total of 26.67 g (682 mmol) of potassium metal, freshly rinsed with 60 mL of dry THF, is added piecewise (Note 7). The reaction mixture is heated at reflux with stirring for 3 hr, at which time the activated magnesium has formed as a finely divided black powder (Note 8). The activated magnesium is allowed to cool and settle. The supernatant THF layer is carefully transferred via cannula into a 500-mL Erlenmeyer flask containing 250 mL of isopropyl alcohol. The activated magnesium is rinsed with dry diethyl ether (2 x 200-mL) and diluted with 100 mL of dry diethyl ether all via cannula. The crude 1-chloro-1-[(dimethyl)phenylsilyl]hexane (31.58 g, 124 mmol) is diluted with 100 mL of dry diethyl ether and transferred via cannula to the addition funnel. The ethereal solution of the α-silyl chloride is added to the activated magnesium slurry in diethyl ether slowly in portions causing the reaction mixture to reflux gently (Note 9). Upon completion of the addition, the reaction mixture is stirred for 15 min (Note 10).

An oven-dried, 1-L, round-bottomed flask is equipped with a magnetic stirring bar and a rubber septum. The flask is charged with 25.61 g of copper bromide-dimethyl sulfide complex (124 mmol) and 250 mL of dry diethyl ether, and the resulting slurry is cooled to -78°C with a dry ice-acetone bath. The Grignard reagent solution prepared above is added to the copper bromide-dimethyl sulfide slurry via cannula

(Note 11). The residual activated magnesium is rinsed once with 200 mL of dry diethyl ether, and the supernatant layer is transferred via cannula to the copper bromide-dimethyl sulfide slurry (Note 12). The reaction mixture is slowly warmed to -10°C, and then 16.78 g (17.42 mL, 125 mmol) of hexanoyl chloride is added dropwise via syringe after which the reaction mixture is warmed to room temperature. After stirring for 3 hr, the reaction mixture is filtered through a 75-g layer of Celite 545 (Note 13) and the filter cake rinsed with three 100-mL portions of diethyl ether. Concentration of the filtrate under reduced pressure yields the α-silyl ketone which is utilized without further purification (Note 14).

(1) Acidic Elimination:

An oven-dried, 1-L, three-necked, round-bottomed flask is equipped with a 125-mL pressure-equalizing addition funnel, a magnetic stirring bar, and a rubber septum, and the system is purged with argon. The flask is charged with the crude α-silyl ketone prepared above and 500 mL of dry THF. Via cannula, a 125-mL (175 mmol) portion of a 1.4 M solution of methyllithium in hexanes is transferred to the addition funnel . After the α-silyl ketone solution is cooled to -78°C with a dry ice-acetone bath, the methyllithium solution is added dropwise over approximately 1 hr, and the reaction mixture is stirred for 0.5 hr at -78°C. A second, oven-dried, 1-L, round-bottomed flask is equipped with a magnetic stirring bar and a rubber septum. The second flask is charged with 47.39 g (249 mmol) of p-toluenesulfonic acid monohydrate and 100 mL of dry THF, and purged with argon. The β-alkoxysilane solution prepared above is transferred to the flask containing the solution of p-toluenesulfonic acid monohydrate in THF via cannula and stirred for 2 hr. The reaction mixture is then poured into a separatory funnel containing a biphasic mixture of 500 mL of saturated aqueous sodium bicarbonate and 250 mL of diethyl ether. The resulting organic layer is washed with saturated aqueous sodium bicarbonate (3 x 250-mL) and brine (250 mL), dried over MgSO$_4$, and concentrated under reduced pressure. Purification of the residue via flash column chromatography (silica gel 230-400 mesh, 450 g, eluant hexanes) yields 11.31-11.75 g of (Z)-6-methyl-6-dodecene (50-52% from the α-silyl chloride) as a 92:8 Z/E ratio of isomers (Notes 15 and 16).

(2) Basic Elimination:

An oven-dried, 1-L, three-necked flask is equipped with a 125-mL pressure-equalizing addition funnel, a magnetic stirring bar, and a rubber septum. The flask is purged with argon and is charged with the crude α-silyl ketone (from 30.00 g, 118 mmol of α-silyl chloride) and dry THF (500 mL). Methyllithium (1.4 M, 118 mL, 165 mmol) is added to the addition funnel via cannula. The α-silyl ketone solution is cooled to -78°C with a dry ice-acetone bath, and methyllithium is added dropwise over approximately 1 hr after which the reaction mixture is stirred for 0.5 hr. A second, oven-dried, 1-L, round-bottomed flask, equipped with a magnetic stirring bar and a rubber septum, is charged with 27 g of potassium hydride (35% wt/wt dispersion in oil which is rinsed with three 100-mL portions of dry diethyl ether), and diluted with 100 mL of dry THF, and 0.200 g of 18-crown-6 (0.76 mmol) is added. The β-alkoxysilane solution prepared above is added to the potassium hydride slurry via cannula, and the mixture is stirred for 16 hr. Excess potassium hydride is quenched with isopropyl alcohol (~ 50 mL) until no further hydrogen gas is evolved (Note 17). Saturated aqueous ammonium chloride (250-mL) is added to the reaction mixture which is then combined with 250 mL of diethyl ether in a separatory funnel. The organic layer is washed successively with three 250-mL portions of saturated aqueous ammonium chloride and 250 mL of brine, dried over MgSO$_4$, and concentrated under reduced pressure. Purification of the residue via flash column chromatography (silica gel 230-400 mesh, 450 g, eluant hexanes) provides 11.19-11.62 g (E)-6-methyl-6-dodecene, (52-54% from the α-silyl chloride) as a 95:5 E/Z ratio of isomers (Notes 15 and 18).

2. Notes

1. Both THF and diethyl ether were obtained from Mallinckrodt Inc. Before use they were dried by distillation from sodium metal and benzophenone under an atmosphere of nitrogen.

2. Lithium wire (3.2 mm diam.), carbon tetrachloride, triphenylphosphine, $MgBr \cdot Et_2O$, copper bromide-dimethyl sulfide complex, hexanoyl chloride, methyllithium, p-toluenesulfonic acid monohydrate, potassium hydride, and 18-crown-6 were purchased from Aldrich Chemical Company, Inc. and used without further purification.

3. The CryoCool bath may be obtained from CryoCool CC-80II Neslab Instruments, Inc. Portsmouth, N.H. 03801, USA. The reaction may be run in an ice bath under supervision.

4. Phenyldimethylsilyl chloride was purchased from the Petrarch Chemical Company and used without further purification.

5. Hexanal was purchased from the Aldrich Chemical Company Inc. and distilled (bp 131°C) before use.

6. The product exhibits the following properties: IR (film) cm^{-1}: 3016, 2968, 2940, 2866, 1465, 1430, 1253, 1117; ^{1}H NMR (400 MHz, CDCl$_3$) δ: 0.40 (s, 3 H), 0.41 (s, 3 H), 0.85 (t, 3 H, J = 7.1), 1.17-1.34 (m, 5 H), 1.58-1.68 (m, 3 H), 3.42 (dd, 1 H, J = 2.9, 11.2), 7.35-7.39 (m, 3 H), 7.54-7.56 (m, 2 H); ^{13}C NMR (75 MHz, CDCl$_3$) δ: -5.7, -4.6, 14.0, 22.5, 27.4, 31.0, 33.1, 51.2, 127.8, 129.5, 134.1, 136.0; high resolution mass spectrum (CI, NH$_3$) m/z 272.1615 [(M$^+$NH$_4$)$^+$; calcd for $C_{14}H_{27}NSiCl$: 272.1602].

7. Potassium metal, purified, was purchased from J.T. Baker Chemical Company. *Caution: Potassium metal is pyrophoric and reacts violently with water.*

8. Activated magnesium is prepared as described in *Org. Synth.* **1979**, *59*, 85.

9. *Caution: The reaction is exothermic.*

10. The Grignard solution may be stored overnight under argon.

11. Care should be taken not to add excess activated magnesium, although a small amount does not seem to affect cuprate formation.

12. Activated magnesium is quenched by the addition of isopropyl alcohol until hydrogen gas is no longer evolved. *Caution*: *Hydrogen gas is flammable and should be vented into a fume hood.*

13. The Celite used is NOT the acid-washed reagent. Acid-washed Celite will cause some desilylation of the α-silyl ketone intermediate.

14. The α-silyl ketone may be stored overnight either under vacuum or under argon at -20°C.

15. The ratio of isomers was determined by GC/MS (Hewlett Packard 5890 Series II Gas Chromatograph/5870 Series Mass Selective Detector).

16. The product exhibits the following properties: IR (film) cm^{-1}: 2966, 2935, 2867, 1460, 1379; ^{1}H NMR (500 MHz, CDCl$_3$) δ: 0.89 (t, 3 H, J = 7.1, overlapping 0.88 (t, 3 H, J = 6.9)), 1.23-1.39 (m, 12 H), 1.58 (s, 0.26 H, E-methyl), 1.67 (s, 2.74 H, Z-methyl), 1.94-2.01 (m, 4 H), 5.11 (t, 1 H, J = 7.1); ^{13}C NMR (125 MHz, CDCl$_3$) δ: 14.10, 22.63, 22.64, 23.40, 27.75, 27.77, 29.80, 31.60, 31.70, 31.80, 125.30, 135.40; high resolution mass spectrum (CI, NH$_3$) m/z 182.2029 [(M)$^+$; calcd for C$_{13}$H$_{26}$: 182.2035].

17. *Caution*: *Hydrogen gas is flammable and should be vented into a fume hood.*

18. The product exhibits the following properties: IR (film) cm^{-1}: 2968, 2936, 2867, 1460, 1380; ^{1}H NMR (500 MHz, CDCl$_3$) δ: 0.89 (t, 6 H, J = 7.0), 1.21-1.41 (m, 12 H), 1.58 (s, 2.73 H, E-methyl), 1.67 (s, 0.27 H, Z-methyl), 1.94-1.99 (m, 4 H), 5.11 (t, 1 H, J = 7.1); ^{13}C NMR (125 MHz, CDCl$_3$) δ: 14.1, 15.8, 22.6, 27.7, 27.9, 29.6, 29.9, 31.6, 31.7, 31.8, 39.7, 124.6, 135.1; high resolution mass spectrum (CI, NH$_3$) m/z 182.2026 [(M)$^+$; calcd for C$_{13}$H$_{26}$: 182.2035].

Waste Disposal Information

All toxic materials were disposed of in accordance with "Prudent Practices for Disposal of Chemicals from Laboratories"; National Academy Press; Washington, DC, 1983.

3. Discussion

The acid or base elimination of a diastereoisomerically pure β-hydroxysilane, **1**, (the Peterson olefination reaction[4]) provides one of the very best methods for the stereoselective formation of alkenes. Either the E- or Z-isomer may be prepared with excellent geometric selectivity from a single precursor (Scheme 1). The widespread use of the Peterson olefination reaction in synthesis has been limited, however, by the fact that there are few experimentally simple methods available for the formation of diastereoisomerically pure β-hydroxysilanes.[5,6] One reliable route is the Cram controlled addition of nucleophiles to α-silyl ketones,[6] but such an approach is complicated by difficulties in the preparation of (α-silylalkyl)lithium species or the corresponding Grignard reagents. These difficulties have been resolved by the development of a simple method for the preparation and reductive acylation of (α-chloroalkyl)silanes.[7]

Scheme 1

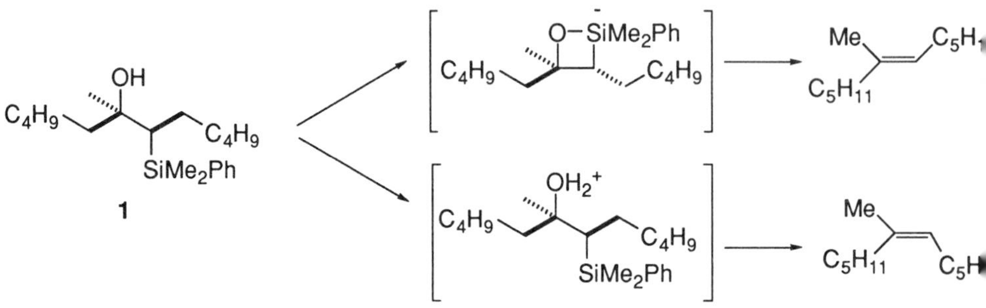

The procedure shown here describes the preparation of α-silyl ketones from aldehydes and acyl chlorides. The α-silyl ketones undergo Cram addition of various nucleophiles to produce diastereoselectively β-hydroxysilanes. These compounds are then subjected directly to elimination in situ under basic or acidic conditions to produce the corresponding alkenes.

1. Department of Chemistry, Colorado State University, Fort Collins, Colorado, USA, 80523.

2. Fleming, I.; Newton, T.W.; Roessler, F. *J. Chem. Soc., Perkin Trans. 1* **1981**, 2527.

3. Appel, R. *Angew Chem., Int. Ed. Engl.* **1975**, *14*, 801.

4. (a) Peterson, D. J. *J. Org. Chem.* **1968**, *33*, 780. For reviews see: (b) Ager, D. J. *Org. React.* **1990**, *38*, 1-223; (c) Ager, D. J. *Synthesis* **1984**, 384; (d) Colvin, E.W. "Silicon Reagents in Organic Synthesis"; Academic Press: New York; 1988; (e) Chan, T.-H. *Acc. Chem. Res.* **1977**, *10*, 442; (f) Hudrlik, P. F.; Peterson, D. J. *J. Am. Chem. Soc.* **1975**, *97*, 1464.

5. (a) Hudrlik, P. F.; Hudrlik, A. M.; Misra, R. N.; Peterson, D.; Withers, G. P.; Kulkarni, A. K. *J. Org. Chem.* **1980**, *45*, 4444; (b) Hudrlik, P. F.; Peterson, D.; Rona, R. J. *J. Org. Chem.* **1975**, *40*, 2263; (c) Sato, F.; Suzuki, Y.; Sato, M. *Tetrahedron Lett.* **1982**, *23*, 4589; (d) Sato, F.; Uchiyama, H.; Iida, K; Kobayashi, Y.; Sato, M. *J. Chem. Soc., Chem. Commun.* **1983**, 921. Also see: (e) Sato, T.; Abe, T.; Kuwajima, I. *Tetrahedron Lett.* **1978**, 259; (f) Yamamoto, K.; Tomo, Y.; Suzuki, S. *Tetrahedron Lett.* **1980**, *21*, 2861; (g) Hudrlik, P. F.; Kulkarni, A. K. *J. Am. Chem. Soc.* **1981**, *103*, 6251; (h) Sato, T.; Matsumoto, K.; Abe, T.; Kuwajima, I. *Bull. Chem. Soc. Jpn.* **1984**, *57*, 2167.

6. (a) Hudrlik, P. K.; Peterson, D. *Tetrahedron Lett.* **1972**, 1785; (b) Utimoto, K.; Obayashi, M.; Nozaki, H. *J. Org. Chem.* **1976**, *41*, 2940.

7. Barrett, A. G. M.; Hill, J. M.; Wallace, E. M. *J. Org. Chem.* **1992**, *57*, 386.

Appendix
Chemical Abstracts Nomenclature (Collective Index Number); (Registry Number)

1-Chloro-1-[(dimethyl)phenylsilyl]hexane: Silane, (1-chlorohexyl)dimethylphenyl- (12); (135987-51-8)

Phenyldimethylsilyl chloride: ALDRICH: Chlorodimethylphenylsilane: Silane, chlorodimethylphenyl- (8,9); (768-33-2)

Hexanal (8, 9); (66-25-1)

1-[(Dimethyl)phenylsilyl]-1-hexanol: 1-Hexanol, 1-(dimethylphenylsilyl)- (12); (125950-71-2)

Carbon tetrachloride, CANCER SUSPECT AGENT (8); Methane, tetrachloro- (9); (56-23-5)

Triphenylphosphine: Phosphine, triphenyl- (8,9); (603-35-0)

(E)-6-Methyl-6-dodecene: 6-Dodecene, 6-methyl, (E)- (11); (101146-61-6)

(Z)-6-Methyl-6-dodecene: 6-Dodecene, 6-methyl-, (Z)- (11); (101165-44-0)

Magnesium bromide etherate: Magnesium, dibromo(ethyl ether) (8); Magnesium, dibromo[1,1'-oxybis[ethane]]- (9); (29858-07-9)

Copper(I) bromide-dimethyl sulfide complex: Copper, bromo[thiobis[methane]]- (9); (54678-23-8)

Hexanoyl chloride (8,9); (142-61-0)

Methyllithium: Lithium, methyl- (8,9); (917-54-4)

p-Toluenesulfonic acid monohydrate (8); Benzenesulfonic acid, 4-methyl-, monohydrate (9); (6192-52-5)

18-Crown-6: 1,4,7,10,13,16-Hexanoxacyclooctadecane: (8,9); (17455-13-9)

SYNTHESIS OF β-LACTONES AND ALKENES VIA THIOL ESTERS:
(E)-2,3-DIMETHYL-3-DODECENE

Submitted by Rick L. Danheiser, James S. Nowick, Janette H. Lee, Raymond F. Miller, and Alexandre H. Huboux.[1]

Checked by David J. Mathre and Ichiro Shinkai.

1. Procedure

A 1-L, three-necked, round-bottomed flask is equipped with a mechanical stirrer, nitrogen inlet adapter, and a 150-mL, pressure-equalizing dropping funnel fitted with a rubber septum (Note 1). The flask is charged with 225 mL of dry tetrahydrofuran (THF) (Note 2) and 26.5 mL (0.189 mol) of diisopropylamine (Note 3), and then is cooled in an ice-water bath while 70.4 mL (0.178 mol) of a 2.53 M solution of n-butyllithium in hexane (Note 4) is added dropwise over 5-10 min. After 10 min, the resulting solution is cooled to -78°C (Note 5) in a dry ice-acetone bath and a solution of 45.0 g (0.17 mol) of S-phenyl decanoate (Note 6) in 75 mL of dry tetrahydrofuran is added dropwise over 1 hr (the dropping funnel is washed with two 5-mL portions of tetrahydrofuran). The yellow reaction mixture is allowed to stir for 30 min at -78°C, and then 18.2 mL (0.17 mol) of 3-methyl-2-butanone (Note 7) is added via a syringe pump or funnel over 7.5 min (Note 8). After 30 min, the reaction mixture is allowed to warm gradually to 0°C over 1.5 hr (Note 9) and then is quenched by the addition of 225 mL of a half-saturated aqueous ammonium chloride solution. The resulting mixture is poured into a 1-L separatory funnel containing 150 mL of hexane and 150 mL of water. The aqueous layer is separated and washed with 50 mL of hexane. The combined organic layers are washed successively with three 200-mL portions of aqueous 10% sodium carbonate and 200 mL of saturated sodium chloride (NaCl) solution, dried over anhydrous sodium sulfate, filtered, and concentrated at reduced pressure using a rotary evaporator to afford 41.5 g of the β-lactone as a yellow oil which is used in the next step without further purification (Note 10).

A 500-mL, one-necked, round-bottomed flask, equipped with a magnetic stirring bar and a condenser fitted with a nitrogen inlet adapter, is charged with the 41.5 g of crude β-lactone prepared above, 200 mL of cyclohexane (Note 11), and 41.5 g of 230-400 mesh silica gel (Note 12). The resulting yellow mixture is heated at reflux for 1 hr (water may be observed collecting in the condenser) and then allowed to cool to room temperature. Activated charcoal (6 g) is added (Note 13), and the resulting mixture is stirred for 5 min and filtered. The residue is washed with an additional 50 mL of cyclohexane and the combined filtrates are concentrated at reduced pressure using a rotary evaporator to afford 31.5 g of a yellow oil. This crude material is applied to the top of a 4.8-cm x 30-cm column of 210 g of 230-400 mesh silica gel 60 (Note 14) and eluted with hexane (20 mL per min, 120-mL fractions) (Note 15). Concentration of fractions 3-5 using a rotary evaporator and then high vacuum (0.1 mm) affords 22.0-23.0 g (66-69% overall yield) of (E)-2,3-dimethyl-3-dodecene as a clear, colorless oil (Notes 16-18).

2. Notes

1. The apparatus is oven-dried at 110°C or flame-dried under reduced pressure and maintained under an atmosphere of nitrogen during the course of the reaction.

2. Tetrahydrofuran was distilled from sodium benzophenone ketyl immediately before use. The checkers used tetrahydrofuran (E. Merck) dried over 4 Å molecular sieves and purged with dry nitrogen (water content <10 µg/mL by Karl Fischer titration).

3. Diisopropylamine was purchased from Aldrich Chemical Company, Inc., and distilled from calcium hydride prior to use. The checkers used diisopropylamine (Aldrich Chemical Company, Inc.) dried over 3 Å molecular sieves and purged with dry nitrogen (water content <35 µg/mL by Karl Fisher titration).

4. n-Butyllithium was purchased from Aldrich Chemical Company, Inc., and titrated according to the the method of Watson and Eastham.[2]

5. The checkers monitored the internal temperature of the reaction with a Teflon-coated J-type thermocouple.

6. S-Phenyl decanoate was prepared by the following procedure:[3] A 500-mL, three-necked, round-bottomed flask equipped with a magnetic stirring bar, glass stopper, nitrogen inlet adapter, and a 50-mL pressure-equalizing addition funnel fitted with a rubber septum is charged with 250 mL of methylene chloride, 17.6 mL (0.171 mol) of thiophenol, and 13.9 mL (0.172 mol) of pyridine (Note 19). The reaction mixture is cooled in an ice-water bath and 35.4 mL (0.171 mol) of decanoyl chloride (Note 19) is added dropwise via the addition funnel over 20 min. The resulting suspension of white solid is stirred for an additional 10 min at 0°C, at room temperature for 1 hr, and then is poured into 210 mL of water. The organic phase is separated and washed, successively with 210 mL of 10% hydrochloric acid and 210

mL of saturated sodium chloride solution, dried over anhydrous magnesium sulfate, filtered, and concentrated at reduced pressure using a rotary evaporator, then under high vacuum (0.1 mm), to provide 44.8 g of a clear, colorless oil. Distillation of this material through an 8-cm Vigreux column affords 41.7-44.0 g (93-98%) of S-phenyl decanoate as a clear, colorless oil, bp 95-125°C (0.04 mm). [Note: it appears that some of the product may decompose during the distillation based on capillary GC analysis (crude product, distilled fractions, and pot residue) and the wide temperature range for the distillation even though the pressure appeared to remain constant.]

7. 3-Methyl-2-butanone was purchased from Eastman Chemical Products, Inc. and distilled prior to use. It appeared that water was azeotropically removed during the distillation, and thus a significant fore-run was discarded.

8. The submitters employed an addition funnel and conducted the addition over 7.5 min.

9. This is best accomplished by periodically adding room temperature acetone to the cooling bath over the course of 1.5 hr. More rapid warming results in dramatically reduced yields of β-lactone.

10. A pure sample of the β-lactone was obtained by vacuum distillation through an 8-cm Vigreux column (bp 43-80°C, 0.25 mm) followed by column chromatography on silica gel and exhibited the following spectral characteristics: IR (neat) cm $^{-1}$: 2960, 2930, 2860, 1830, 1465, 1390, 1220, 1095, 1020, 810; ^{1}H NMR (300 MHz, CDCl$_3$) δ: 0.88 (t, 3 H, J = 7), 0.93 (d, 3 H, J = 7), 1.02 (d, 3 H, J = 7), 1.20-1.50 (m, 15 H), 1.50-1.65 (m, 1 H), 1.65-1.90 (m, 1 H), 1.99 (sept, 1 H, J = 7), 3.13 (t, 1 H, J = 8); ^{13}C NMR 75 MHz, CDCl$_3$) δ: 14.0, 14.9, 17.0, 17.5, 22.6, 25.0, 27.5, 29.1, 29.2, 29.4, 31.8, 37.5, 56.2, 85.0,171.9. Anal. Calcd for C$_{15}$H$_{28}$O$_2$: C, 74.95; H, 11.74. Found: C, 75.17; H, 11.57. Only the trans-substituted β-lactone could be detected by NMR and NOE analysis.

11. Cyclohexane was purchased from Mallinckrodt Inc. and used without further purification.

12. Silica gel (230-400 mesh) was obtained from J. T. Baker, Inc., or E. Merck. Less silica gel (e.g., 10% wt. equiv) can be used in this step, although in this case longer reflux times are necessary to complete the decarboxylation. Note that the use of completely anhydrous silica gel (e.g., silica gel flame-dried under reduced pressure) was found to catalyze olefin isomerization in some cases and should be avoided.

13. Activated charcoal (20-40 mesh) was used as received from Matheson, Coleman, & Bell or Darco (G-60). Addition of charcoal at this stage removes reduces the amount of malodorous thiol impurities.

14. The submitters silica gel (J. T. Baker, Inc.) column (8-cm x 10-cm) was packed as a slurry in petroleum ether (Mallinkrodt Inc., bp 35-60°C). The column was eluted using petroleum ether (20 mL/min, collecting 70-mL fractions). The fractions were analyzed by capillary GC, and those containing the product (fractions 4-9) were combined.

15. Elution of the silica gel column with additional hexane afforded thiophenol (5.0 g), diphenyl disulfide (0.2 g), and BHT (0.1 g, stabilizer from the THF). Continued elution with increasing amounts of ethyl acetate (hexane:EtOAc from 100:0 to 80:20) afforded dinonyl ketone (2.4 g, 10%), S-phenyl decanoate (0.8 g, 2%), and the enol of 2-methyl-3,5-diketotetradecane (2.4 g, 6%; the amount is probably greater).

16. Alternatively, the alkene can be purified by distillation through an 8-cm Vigreux column (Note 20) to furnish 11.8-13.0 g (53-59% overall yield) of (E)-2,3-dimethyl-3-dodecene as a clear, colorless oil, bp 52°C (0.03 mm) The yield of alkene is considerably reduced if the distillation is carried out at higher temperature.

17. The olefin thus obtained was found by ^{1}H NMR and gas chromatographic analysis to consist of a 25-27:1 mixture of E and Z isomers. The major product was identified as the trans isomer by ^{1}H NMR NOE analysis. Gas chromatographic

analysis was carried out on a 0.25-mm x 30-m DB-1701-coated fused silica capillary column, column temperature 125°C, flow rate 1 mL/min, retention times: Z isomer 9.09 min, E isomer 9.30 min.

18. The product has the following spectral properties: IR (neat) cm^{-1}: 2970, 2940, 2870, 1465, 1380; ^{1}H NMR (300 MHz, CDCl$_3$) δ: 0.87 (t (br), 3 H), 0.95 (d, 6 H, J= 6.5), 1.25 (s (br), 12 H), 1.55 (s, 3 H), 1.95 (m, 2 H), 2.20 (m, 1 H), 5.15 (t (br), 1 H, J = 6.9); ^{13}C NMR (75 MHz, CDCl$_3$) δ: 13.3, 14.1, 21.5, 22.7, 27.8, 29.4, 29.6, 30.0, 32.0, 36.8, 122.3, 140.6; Z isomer impurity (partial) δ: 5.05 (t (br), J = 6.5 Hz), 2.8 (m), 1.60 (m), 0.95 (d); Anal. Calcd for C$_{14}$H$_{28}$: C, 85.71; H, 14.29. Found: C, 85.55; H, 14.38. The major product was determined to be the trans isomer by NOE analysis.

19. Methylene chloride was purchased from Mallinckrodt Inc. and used without further purification. Thiophenol was purchased from Fluka Chemical Co. and used without further purification. Pyridine was purchased from Mallinckrodt Inc. and was distilled from calcium hydride. Decanoyl chloride was purchased from the Aldrich Chemical Company, Inc., and used without further purification.

20. In order to reduce foaming, enough glass wool is added to the distillation flask just to cover the surface of the liquid.

Waste Disposal Information

All toxic materials were disposed of in accordance with "Prudent Practices for Disposal of Chemicals from Laboratories"; National Academic Press; Washington, DC, 1983.

3. Discussion

The procedure described here illustrates a practical and convenient method for the synthesis of β-lactones.[3] In conjunction with the stereospecific decarboxylation of the β-lactone products, it provides an attractive strategy for the synthesis of substituted alkenes. The new method is based on the addition of lithium enolate derivatives of thiol esters to ketones and aldehydes at -78°C as originally described by Wemple.[4] The submitters have found that gradual warming (generally to 0°C) of the resulting aldolates produces β-lactones in good to excellent yield. The facility and generality of this spontaneous lactonization process had not been noted previously. Low-temperature quenching of the aldol addition reaction affords only the expected β-hydroxy thiol esters, although Masamune has shown that cyclization of the thiol ester aldol products can be effected as a separate operation by treatment with excess mercury(II) methanesulfonate and disodium phosphate (Na_2HPO_4) in acetonitrile.[5]

Both ketones and aldehydes, as well as acylsilanes can be employed as carbonyl substrates in the new β-lactone synthesis (Table). Reactions involving ketones are most conveniently carried out by adding the neat carbonyl compound to the thiol ester enolate solution. Under these conditions aliphatic aldehydes react to form substantial quantities of 2:1 adducts; however, formation of these side products can be suppressed simply by slowly adding the aldehyde component as a precooled (-78°C) solution to the reaction mixture. Wide variation is also possible in the thiol ester component, although a few limitations of the method have been noted. For example, α,β-unsaturated ketones such as methyl vinyl ketone and cyclohexenone fail to yield β-lactones, and attempts to generate β-lactones with severe steric crowding have also met with limited success.[3]

Not surprisingly, thiol ester enolates combine with ketones (and many aldehydes) to form predominantly the less sterically crowded β-lactone diastereomers,

67

in some cases with excellent stereoselectivity. However, the stereochemical course of reactions involving aldehydes has proved to be rather complicated, and further studies are required to clarify the factors that control the stereochemical outcome of these reactions.

As described here, the new β-lactone synthesis also provides the basis for a very attractive approach to the synthesis of substituted alkenes. Since the 19th century it has been known that, upon heating, β-lactones undergo a facile [2+2] cycloreversion to generate alkenes and carbon dioxide.[6] This stereospecific process[7] generally takes place at temperatures between 80° and 160°C, with the rate of reaction being highly dependent on the nature of substituents present at the C-4 (β) position of the lactone ring. The reaction often proceeds in nearly quantitative yield, and in recent years has been applied to the synthesis of a variety of types of substituted and functionalized alkenes. In this work, two experimental protocols were employed to effect the cycloreversion step. Alkenes boiling at 200°C or lower were best generated by Kugelrohr distillation at 80-110°C from a mixture of the lactone and 10 weight % of silica gel[8] at a pressure such that the alkene product distilled as it was generated, leaving the less volatile β-lactone behind in the distillation flask. As described in the above procedure, alkenes with boiling points of ca. 250°C or greater were prepared by heating a benzene or cyclohexane solution of the requisite β-lactone at reflux in the presence of an equal weight of chromatographic silica gel.

The synthesis of β-lactones has received considerable attention[6] since the first representative of this class of heterocycles was prepared in 1883. Classical routes to β-lactones generally involved the cyclization of β-halocarboxylate salts[6] and the related "deaminative cyclization" that occurs upon diazotization of β-amino acids.[9] β-Hydroxy acids undergo a similar cyclization under Mitsunobu conditions,[10] and the halolactonization of α,β-unsaturated acids[11] is a related process of some interest. Although these classical methods have been successfully employed for the preparation of a variety of β-lactones, their utility is often limited by side reactions including β-elimination (to form α,β-unsaturated acids) and decarboxylative elimination (to generate alkenes).

The strategy described here should find considerable use as a method for the stereoselective synthesis of alkenes. Although this olefination strategy involves one more step than the classic Wittig reaction, in many cases it may prove to be the more practical method. Finally, the scope, overall efficiency, and stereoselectivity of the β-lactone route compares favorably to the Wittig, Julia-Lythgoe, and related established strategies for the synthesis of tri- and tetrasubstituted alkenes.

1. Department of Chemistry, Massachusetts Institute of Technology, Cambridge, MA 02139. We thank the National Science Foundation and the National Institutes of Health for generous financial support. R. F. M. was supported in part by NIH training grant CA 09112.

2. Watson, S. C.; Eastham, J. F. *J. Organomet. Chem.* **1967**, *9*, 165.

3. (a) Danheiser, R. L.; Nowick, J. S. *J. Org. Chem.* **1991**, *56*, 1176; (b) Danhesier, R. L.; Choi, Y. M.; Menichincheri, M.; Stoner, E. J. *J. Org. Chem.* **1993**, *58*, 322.

4. Wemple, J. *Tetrahedron Lett.* **1975**, 3255.

5. Masamune, S.; Hayase, Y.; Chan, W. K.; Sobczak, R. L. *J. Am. Chem. Soc.* **1976**, *98*, 7874.

6. For reviews of the chemistry and synthesis of β-lactones, see: (a) Searles, G. In "Comprehensive Heterocyclic Chemistry"; Katritzky, A.R.; Rees, C. W., Eds.; Pergamon: Oxford, 1984; Vol. 7; Chapter 5.13; pp 363-402; (b) Pommier, A.; Pons, J.-M. *Synthesis* **1993**, 441.

7. Mulzer, J.; Pointher, A.; Chucholowski, A.; Brüntrup, G. *J. Chem. Soc., Chem. Commun.* **1979**, 52.

8. The use of silica gel to promote the decarboxylation of β-lactones has previously been reported by Adam, W.; Encarnacion, L. A. A. *Synthesis* **1979**, 388.

9. Testa, E.; Fontanella, L.; Cristiani, G.; Mariani, L. *Justus Liebigs Ann. Chem.* **1961**, *639*, 166.

10. (a) Mitsunobu, O. *Synthesis*, **1981**, 1; (b) Arnold, L. D.; Drover, J. C. G.; Vederas, J. C. *J. Am. Chem. Soc.* **1987**, *109*, 4649; (c) Adam, W.; Narita, N.; Nishizawa, Y. *J. Am. Chem. Soc.* **1984**, *106*, 1843, and references cited therein.

11. For examples see (a) Barnett, W. E.; Needham, L. L. *J. Org. Chem.* **1975**, *40*, 2843; (b) Holbert, G. W.; Ganem, B. *J. Am. Chem. Soc.* **1978**, *100*, 352.

TABLE I. PREPARATION OF ß-LACTONES VIA THIOL ESTERS

Entry	Carbonyl Compound	Thiol Ester	ß-Lactone	%Isolated Yield
1	Cyclohexanone	CH₃COSPh		86
2	Cyclohexanone	CH₃CH₂COSPh		92
3	Cyclohexanone	MeOCH₂COSPh		78
4	n-Octanal	Me₂CHCOSPh		54
5	CH₃COSi-t-BuMe₂	CH₃COSPh		55
6	Acetylcyclohexane	CH₃CH₂COSPh	29:1	75
7	n-Octanal	CH₃CH₂COSPh	2.5:1	42
8	Benzaldehyde	t-BuCH₂COSPh	34:1	85

71

Appendix
Chemical Abstracts Nomenclature (Collective Index Number); (Registry Number)

Diisopropylamine (8); 2-Proparamine, N-(1-methylethyl)- (9); (108-18-9)

Butyllithium: Lithium, butyl- (8,9); (109-72-8)

2-Phenyl decanoate: Decanethioic acid, S-phenyl ester (9); (51892-25-2)

3-Methyl-2-butanone: TOXIC: 2-Butanone, 3-methyl- (8,9); (563-80-4)

Thiophenol: Benzenethiol (8,9); (108-98-5)

Decanoyl chloride (8,9); (112-13-0)

ALKYLIDENATION OF ESTER CARBONYL GROUPS:
(Z)-1-ETHOXY-1-PHENYL-1-HEXENE

A. CH_2Br_2 $\xrightarrow[\text{2) n-C}_4\text{H}_9\text{I, HMPA}]{\substack{\text{1) LDA, THF-ether-hexane} \\ -90°C}}$ $n\text{-}C_4H_9CHBr_2$

B. $+$ $n\text{-}C_4H_9CHBr_2$ $\xrightarrow[\text{THF, 25°C}]{\substack{\text{TiCl}_4\text{, TMEDA} \\ \text{Zn, cat. PbCl}_2}}$

Submitted by Kazuhiko Takai,[1] Yasutaka Kataoka, Jiro Miyai, Takashi Okazoe, Koichiro Oshima, and Kiitiro Utimoto.[2]

Checked by Stephen T. Wrobleski, Alan T. Johnson and Robert K. Boeckman, Jr.

1.   Procedure

Caution! Hexamethylphosphoric triamide (HMPA) is toxic and must be handled with gloves.

A.   *1,1-Dibromopentane* (Note 1).   A dry, 3-L, three-necked, round-bottomed flask is equipped with a mechanical stirrer, nitrogen inlet, rubber septum, and a 200-mL, graduated, pressure-equalizing addition funnel that is sealed with a rubber septum. After placing the system under a nitrogen atmosphere, the flask is charged with 280 mL of dry tetrahydrofuran (Note 2), 400 mL of dry diethyl ether (Note 2), and 20 g (0.20 mol) of dry diisopropylamine (Note 3). Stirring is initiated, and the contents of the flask are cooled to -10°C (dry ice-acetone bath). After transfer of 120 mL (0.20

mol) of 1.7 M butyllithium in hexane (Note 4) to the addition funnel by syringe, the solution of alkyllithium is slowly added to the stirred solution at such a rate as to maintain a temperature of -10°C. After the addition is complete, 10 mL of dry THF is introduced via syringe to rinse the walls of the addition funnel and then added to the reaction mixture. The mixture is stirred at -10°C for 15 min and cooled to -90°C (toluene-liq. nitrogen bath). The addition funnel is charged by syringe with a solution of 15 mL (38 g, 0.22 mol) of dibromomethane (Note 5) in 100 mL of dry THF, which is then added at such a rate that the temperature does not exceed -85°C (Note 6). The mixture is stirred for 15 min. A solution of 22 mL (37 g, 0.20 mol) of 1-iodobutane (Note 7) in 100 mL of THF, and 120 mL of hexamethylphosphoric triamide (Note 8) are successively added by syringe at such a rate (over an ~25 min period) that the temperature of the reaction mixture does not exceed -85°C (Note 6). After the addition is complete, the reaction mixture is stirred at -90°C (toluene-liq. nitrogen bath) for 2 hr, -78°C (dry ice-acetone bath) for 1 hr, -48°C (dry ice-acetonitrile bath) for 2 hr, and -23°C (dry ice-carbon tetrachloride bath) for 3 hr (Notes 6 and 9). The solution is poured into 200 mL of 1 M hydrochloric acid solution and transferred to a 2-L separatory funnel. The resulting mixture is extracted with three 100-mL portions of hexane. The organic extracts are combined and washed with three 300-mL portions of water, 100 mL of saturated aqueous sodium sulfite solution, and 100 mL of brine. The organic layer is dried over anhydrous magnesium sulfate and then concentrated with a rotary evaporator (0°C water bath, trapping at -78°C with a dry ice-acetone condenser) at aspirator pressure. The brown liquid residue is distilled to afford 27-32 g (59-70%) of 1,1-dibromopentane as a colorless liquid, bp 70-72°C (15 mm) (Note 10).

   B. *(Z)-1-Ethoxy-1-phenyl-1-hexene.* A dry, 3-L, three-necked, round-bottomed flask is equipped with a mechanical stirrer, nitrogen inlet, rubber septum, and a 200-mL pressure-equalizing addition funnel that is sealed with rubber septum. After placing the system under a nitrogen atmosphere, the flask is charged with 350 mL of dry tetrahydrofuran (Note 2). The contents of the flask are cooled to 0°C with an ice bath, and 140 mL (0.28 mol) of a 2.0 M solution of titanium tetrachloride in dichloromethane (Note 11) is added slowly by syringe to the stirred THF over a period of 10 min. To the yellow solution at 0°C is added slowly 84 mL (0.56 mol) of tetramethylethylenediamine (Note 12) by syringe. After being stirred at 0°C for 20 min, 41 g (0.63 mol) of zinc dust (Note 13) is added to the reaction mixture at 0°C in five portions in such a manner that the temperature remains at 0°C (Note 14), followed by addition of 0.88 g (3.2 mmol) of lead (II) chloride (Notes 13 and 15), and then the resulting suspension is warmed to 25°C. The color of the suspension turns from brownish yellow to dark greenish blue while being stirred at 25°C for 30 min. The addition funnel is then charged by syringe with a solution of 11 g (70 mmol) of ethyl benzoate (Note 12) and 35 g (0.15 mol) of 1,1-dibromopentane (part A) in 100 mL of dry THF. The resulting solution is then added to the stirred reaction mixture over a period of 10 min at 25°C and stirring is continued for 3.5-4.5 hr. The color of the resulting mixture gradually turns dark brown as the reaction proceeds (Notes 16-18). The reaction mixture is then cooled to 0°C and 70 mL of triethylamine (Note 12) and 91 mL of saturated aqueous potassium carbonate solution are successively added slowly at 0°C by syringe. After stirring at 0°C for an additional 15 min, 200 mL of ether-triethylamine (200/1, v/v) is added to the reaction mixture. The entire reaction mixture is then passed rapidly through a thin pad of activity III basic alumina (Note 19) on a 1-L glass filter using 500 mL of ether-triethylamine (200/1, v/v) as eluent. The filtrate is concentrated with a rotary evaporator (25°C, water bath). If a white solid appears at this point, the mixture is diluted with 100 mL of hexane-triethylamine (200/1, v/v) and the mixture is again filtered through a thin pad of basic alumina (Akt. III). The pad is washed with 100 mL of hexane-triethylamine (200/1, v/v) and the total eluent is concentrated again with a rotary evaporator. The resulting crude material is then vacuum distilled to give 11.0-11.4 g (77-80%) of a 93:7 mixture of (Z) and (E)-1-ethoxy-1-phenyl-1-hexene, bp 73-75°C (0.20 mm) (Notes 20 and 21).

2. Notes

1. This procedure was reported by J. Villieras, C. Bacquet, and J. F. Normant.[3]

2. Tetrahydrofuran and diethyl ether were distilled from sodium and benzophenone just before use.

3. Diisopropylamine was distilled from calcium hydride, bp 84°C.

4. A 1.7 M hexane solution of butyllithium was obtained from Kanto Chemical Co. It may be standardized; however, the submitters chose to use a fresh reagent and forego the titration. The checkers employed a 1.6 M solution of n-butyllithium in hexane obtained from Lithco Inc., which was standardized before use.

5. Dibromomethane was freshly distilled, bp 96-97°C. The checkers noted some variability in the yield which in part appeared to be associated with the source of the dibromomethane.

6. The checkers monitored the internal temperature of the reaction mixture via thermocouple using an immersion well. The checkers observed that accurate temperature control is essential to obtain the reported yields reproducibly.

7. 1-Iodobutane was freshly distilled, bp 129°C.

8. Hexamethylphosphoric triamide is toxic and a cancer-suspect agent. It was distilled from calcium hydride, bp 68-69°C at 1 mm.

9. The color of the mixture changed from brown to white after 30-min stirring at -90°C. Then the color of the mixture changed gradually from white to light brown at -23°C.

10. The infrared spectrum (neat) has absorptions at 2956, 2930, 2860 1465, 1431, 1238, 1158, 927, 732, 667, 596 cm[-1]; ^{1}H NMR (CDCl$_3$) δ: 0.95 (t, 3 H, J = 7.2),

1.28-1.61 (m, 4 H) 2.41 (dt, 2 H, J = 8.4, 6.2), 5.72 (t, 1 H, J = 6.2); ^{13}C NMR (CDCl$_3$) δ: 13.8, 21.4, 30.1, 45.1, 46.2.

11. Freshly distilled titanium tetrachloride (bp 136°C) was diluted with dichloromethane to afford a 2.0 M solution. All residues of titanium tetrachloride were destroyed with acetone from a wash bottle.

12. Tetramethylethylenediamine was freshly distilled from potassium hydroxide, bp 46-47°C (47 mm). Ethyl benzoate was distilled before use. Triethylamine was distilled from potassium hydroxide before use.

13. Zinc dust purchased from Wako Pure Chemical Industries, Ltd. (GR grade) was activated by washing several times with 5% hydrochloric acid washing in turn with water, methanol, and ether, and drying in vacuo according to Fieser and Fieser.[4] The lots employed by the submitters were found to contain ~0.05 mol% Pb based on the Zn content by X-ray fluorescence analysis (Note 15). The checkers employed Zn dust (-325 mesh, 99.998% purity) obtained from Aldrich Chemical Company, Inc.

14. The reduction is a slightly exothermic process.

15. Addition of a catalytic amount of PbCl$_2$ (Rare Metallic Co., 99.999% purity) to a commercial lot of Zn dust (Aldrich Chemical Co. (99.998% purity) or Rare Metallic Co. (99.999% purity)) has shown reproducible results as were previously reported.[5,6] The yield of (Z) and (E)-1-ethoxy-1-phenyl-1-hexene fell to 10-15% ((Z)/(E) = 92/8-95/5) without the addition of PbCl$_2$.[7]

16. The consumption of ethyl benzoate was checked by tlc analysis.

17. The following ratios of reactants, ester/1,1-dibromo-alkane/zinc/TiCl$_4$/PbCl$_2$/TMEDA = 1/2.2/9/4/0.045/8, gave the best results. When the amount of the reagent was reduced to 1/1.1/4.5/2/0.023/4, only 44% of the 1-ethoxy-1-phenyl-1-hexene was isolated under the same reaction conditions and 44% of ethyl benzoate remained.

18. Under the reaction conditions for alkylidenation, compounds containing the following functional groups were found to be stable: trimethylsilyl ethers of alcohols, olefins, primary alkyl iodides, and ethylene acetals of aldehydes.

19. Basic alumina (ICN alumina B-Act. I) was purchased from ICN Biochemical GmbH and pretreated by shaking with 6% of water to change its activity (Act I → III).

20. The infrared spectrum (neat) has absorptions at 2924, 2870, 1649, 1492, 1446, 1266, 1072, 768, 696 cm^{-1}; ^{1}H NMR ((Z) isomer) (CDCl$_3$) δ: 1.00 (t, 3 H, J = 6.9), 1.23-1.45 (m, 7 H), 2.32-2.39 (m, 2 H), 3.76 (q, 2 H, J = 7.0), 5.38 (t, 1 H, J = 7.3), 7.23-7.50 (m, 5 H); ^{1}H NMR ((E) isomer) (CDCl$_3$) δ: 0.87 (t, 3 H, J = 7.0), 1.23-1.45 (m, 7 H), 2.00-2.13 (m, 2 H), 3.80 (q, 2 H, J = 7.0), 4.74 (t, 1 H, J = 7.4), 7.23-7.50 (m, 5 H); ^{13}C NMR ((Z) isomer) (CDCl$_3$) δ: 14.0, 15.3, 22.5, 25.4, 32.0, 65.8, 115.4, 125.8, 127.5, 128.2, 136.7, 153.2. MS m/z (%): 204 (M$^+$, 48), 161 (100), 133 (55), 55 (49). Anal. Calcd for C$_{14}$H$_{20}$O: C, 82.30; H, 9.87. Found: C, 82.47; H, 10.04.

21. The ratio of the geometric isomers of the product ((Z)/(E)) was determined by ^{1}H NMR since isomerization has been shown to occur under GLC conditions.[8]

3. Discussion

The procedure described here provides a convenient method for the conversion of esters to Z-alkenyl ethers.[5] The results in the Table show the wide applicability and high Z selectivity of the process. As the substituents R^1 or R^3 become bigger, or R^2 becomes smaller, higher Z selectivity is observed. The stereochemistry of the isomers (Table, cases 1-10) was determined by ^{13}C NMR.[8] Since isomerization of alkenyl ethers has been reported to take place under GLC conditions, the remaining Z/E ratios were measured by ^{1}H NMR (200 MHz) analysis. Esters having terminal double bonds reacted to afford the corresponding alkenyl ethers in about 50% yield (cases 7 and 9). Esters with internal double bonds gave better yields and the stereochemistry of double bonds of the reactants was retained except in the instance of case 8 where partial isomerization of the isolated *cis* double bond occurred. Thus, the reaction provides a convenient and stereoselective access to allyl vinyl ethers (cases 9 and 10) and oxygen-substituted dienes (case 6). Z-Isomers of silyl enol ethers (cases 11-13) and an alkenyl sulfide (case 14) are also produced under good stereocontrol from the corresponding carboxylic acid derivatives.

The preparation of alkenyl ethers is limited to methods which use as starting materials either acetals[9] or acetylenes.[10,11] It is usually difficult to prepare the alkenyl ethers, especially trisubstituted ones, in a regio- and stereoselective manner by these methods. Alkylidenation of carboxylic acid derivatives does not proceed with the Wittig reagents.[12] Methylene transfer ($C=O \rightarrow C=CH_2$) of such electron-rich carbonyl compounds has been achieved with the Tebbe reagent[13] or dimethyltitanocene.[14,15] Alkylidenation by using the Schrock-type metal carbene complex of Ta, Ti, Zr, or W has been reported.[16] However, the method using the Schrock complexes has several drawbacks. For example, i) the preparation of the complex usually requires special techniques and some restrictions exist on the nature of the substitutents R^1 and/or R^2. ii) The alkylidenation reaction using the Schrock complexes does not provide alkenyl ethers with good control over olefin geometry. The present procedure offers an experimentally simple and stereoselective preparation of alkenyl ethers. The reactants, 1,1-dibromoalkanes, are readily prepared from either iodoalkanes (vide supra) or aldehydes[17] and it is not necessary to isolate the reactive organometallic compound.

1. Department of Applied Chemistry, Faculty of Engineering, Okayama University, Tsushima, Okayama 700, Japan.

2. Division of Material Chemistry, Faculty of Engineering, Kyoto University, Yoshida, Kyoto 606-01, Japan.

3. Villieras, J.; Bacquet, C.; Normant, J.-F. *Bull. Soc. Chim. Fr.* **1975**, 1797.

4. Fieser, L. F.; Fieser, M. "Reagents for Organic Synthesis"; Wiley: New York, 1967; Vol. I, p. 1276.

5. Okazoe, T.; Takai, K.; Oshima, K.; Utimoto, K. *J. Org. Chem.* **1987**, *52*, 4410; Takai, K.; Kataoka, Y.; Okazoe, T.; Utimoto, K. *Tetrahedron Lett.* **1988**, *29*, 1065; Takai, K.; Fujimura, O.; Kataoka, Y.; Utimoto, K. *Tetrahedron Lett.* **1989**, *30*, 211.

6. Takai, K.; Kakiuchi, T.; Kataoka, Y.; Utimoto, K. *J. Org. Chem.* **1994**, *59*, 2668.

7. Professor R. K. Boeckman, Jr. and Dr. A. T. Johnson noted while attempting to check this procedure that this alkylidenation did not proceed with commercial samples of high purity zinc from Aldrich Chemical Company, Inc. However, the reaction proceeded normally with a sample of Wako Pure Chemicals zinc powder obtained from the submitters. Atomic absorption analysis by the checkers suggested the only significant impurities to be lead and cadmium. The presence of lead in active lots of zinc powder was also found by the submitters by X-ray fluorescence analysis (with the aid of Professor H. Takatsuki and Mr. Y. Honda of the Environmental Research Center, Kyoto University), and Auger electron spectroscopy (by Professor Z. Takehara and Dr. Y. Uchimoto of the Division of Energy and Hydrocarbon Chemistry, Faculty of Engineering, Kyoto University).

8. Strobel, M. P.; Andrieu, C. G.; Paquer, D.; Vazeux, M.; Pham, C. C. *Nouv. J. Chim.* **1980**, *4*, 603.

9. Wohl, R. A.; *Synthesis* **1974**, 38; Miller, R. C.; McKean, D. R. *Tetrahedron Lett.* **1982**, *23*, 323.

10. Riediker, M.; Schwartz, J. *J. Am. Chem. Soc.* **1982**, *104*, 5842; Utimoto, K. *Pure Appl. Chem.* **1983**, *55*, 1845; Tamao, K.; Kakui, T.; Kumada, M. *Tetrahedron Lett.* **1980**, *21*, 4105.

11. For an example of a method using other starting materials: Gilbert, J. C.; Weerasooriya, U.; Wiechman, B.; Ho, L. *Tetrahedron Lett.* **1980**, *21*, 5003.

12. Uijttewaal, A. P.; Jonkers, F. L.; van der Gen, A. *J. Org. Chem.* **1979**, *44*, 3157.

13. Pine, S. H.; Pettit, R. J.; Geib, G. D.; Cruz, S. G.; Gallego, C. H.; Tijerina, T.; Pine, F. D. *J. Org. Chem.* **1985**, *50*, 1212.

14. Petasis, N. A.; Bzowej, E. I. *J. Am. Chem. Soc.* **1990**, *112*, 6392.

15. The reagent derived from CH_2Br_2, $TiCl_4$, zinc, (cat. Pb) and TMEDA is also effective for methylenation of esters in some cases. See, Barrett, A. G. M.; Bezuidenhoudt, B. C.; Melcher, L. M. *J. Org. Chem.* **1990**, *55*, 5196.

16. (a) Ta: Schrock, R. R. *J. Am. Chem. Soc.* **1976**, *98*, 5399; (b) Zr: Clift., S. M. Schwartz, J. *J. Am. Chem. Soc.* **1984**, *106*, 8300; (c) W: Aguero, A.; Kress, J.; Osborn, J. A. *J. Chem. Soc., Chem. Commun.* **1986**, 531.

17. For other methods for preparation of 1,1-dibromoalkanes, see: (a) Hawkins, B. L.; Bremser, W.; Borcíc, S.; Roberts, J. D. *J. Am. Chem. Soc.* **1971**, *93*, 4472; (b) Kropp, P. J. ; Pienta, N. J. *J. Org Chem.* **1983**, *48*, 2084; (c) Martínez, A. G.; Fernández, A. H.; Alvarez, R. M.; Fraile, A. G.; Claderón, J. B.; Barcina, J. O. *Synthesis* **1986**, 1076.

ALKYLIDENATION OF CARBOXYLIC ACID DERIVATIVES[a]

$$R^1 \overset{O}{\underset{}{\|}}{-}OR^2 \; + \; R^3CHBr_2 \quad \xrightarrow[\text{THF, 25°C}]{\substack{\text{TiCl}_4, \text{TMEDA} \\ \text{Zn, cat. PbCl}_2}} \quad R^1{-}C(OR^2){=}CH{-}R^3$$

Case	R^1	R^2	R^3	Yield(%)[b]	Z/E[c]
1	Bu	Me	n-C$_5$H$_{11}$	96	91/9
2	Bu	Me	i-Bu	95	93/7
3	Bu	Me	c-C$_6$H$_{11}$[d]	69	93/7
4	i-Bu	Me	n-C$_5$H$_{11}$	88	92/8
5	i-Pr	Me	n-C$_5$H$_{11}$	89	100/0
6	(E)-MeCH=CH	Et	n-C$_5$H$_{11}$	90	94/6
7	CH$_2$=CH(CH$_2$)$_8$	Me	Me[e]	53	89/11
8	n-C$_8$H$_{17}$CH=CH(CH$_2$)$_7$	Me	Me[e]	70[f]	90/10
9	Bu	CH$_2$=CHCH$_2$	n-C$_5$H$_{11}$	52	92/8
10	Pr	PrCH=CHCH$_2$	n-C$_5$H$_{11}$	85	94/6
11	n-C$_5$H$_{11}$	Me$_3$Si	PhCH$_2$	76[g]	89/11
12	PhCH$_2$CH$_2$	Me$_3$Si	Bu	80[g]	84/16
13	PhCH=CH	Me$_3$Si	Bu	79[g]	100/0

| 14 | | PhCH2 | 87[g] | 90/10 |
| 15 | | PhCH2 | 87[h] | --- |

---

[a]The carboxylic acid derivatives (1-2 mmol scale) are treated with 1,1-dibromoalkane (2.2 equiv), Zn (9.0 equiv), TiCl$_4$ (4.0 equiv), PbCl$_2$ (0.045 equiv), and TMEDA (8.0 equiv) in THF at 25°C for 0.5-3 hr. [b]Products are isolated by column chromatography on alumina unless otherwise noted. [c]See the text. [d]Prepared from cyclohexanecarbaldehyde.[17b] [e]Commercially available at Tokyo Kasei Kogyo Co., Japan. [f]The cis/trans ratio of the double bond at C-11 of the resultant ether was 87/13. [g]Isolated by column chromatograpy on silica gel (70-230 mesh ATSM, MERCK). [h]The product was isolated by Kugelrohr distillation.bp 150-152°C (bath temperature, 0.2 mm).

Appendix
Chemical Abstracts Nomenclature (Collective Index Number); (Registry Number)

Hexamethylphosphoric triamide: Phosphoric triamide, hexamethyl- (8,9); (680-31-9)

1,1-Dibromopentane: Pentane, 1,1-dibromo- (9); (13320-56-4)

Diisopropylamine (8); 2-Propanamine, N-(1-methylethyl)- (9); (108-18-9)

Butyllithium: Lithium, butyl- (8, 9); (109-72-8)

Dibromomethane: Methane, dibromo- (8,9); (74-95-3)

1-Iodobutane: Butane, 1-iodo- (8,9); (542-69-8)

Titanium tetrachloride: HIGHLY TOXIC: Titanium chloride (8,9); (7550-45-0)

N,N,N',N'-Tetramethylethylenediamine: Ethylenediamine, N,N,N',N'-tetramethyl- (8);

1,2-Ethanediamine, N,N,N',N'-tetramethyl- (9); (110-18-9)

Zinc (8,9); (7440-66-6)

Lead(II) chloride: Lead chloride (8,9); (7758-95-4)

Ethyl benzoate: Benzoic acid, ethyl ester (8,9); (93-89-0)

Triethylamine (8); Ethanamine, N,N-diethyl- (9); (121-44-8)

SYNTHESIS OF 7-SUBSTITUTED INDOLINES via DIRECTED LITHIATION OF 1-(tert-BUTOXYCARBONYL)INDOLINE: 7-INDOLINECARBOXALDEHYDE

A.
$$\xrightarrow{\text{(Boc)}_2\text{O}}$$

B.
$$\xrightarrow[\substack{2.\ \text{DMF/-78°C} \\ 3.\ \text{aq NH}_4\text{Cl}}]{1.\ \text{sec-BuLi/TMEDA/ether/-78°C}}$$

C.
$$\xrightarrow[\substack{2.\ \text{aq NH}_3}]{1.\ \text{concd HCl}}$$

Submitted by Masatomo Iwao and Tsukasa Kuraishi.[1]
Checked by Jeff Crowley and Stephen F. Martin.

1. Procedure

Caution! Part C of this procedure should be carried out in an efficient hood to avoid exposure to noxious vapors (hydrogen chloride and ammonia).

A. 1-(tert-Butoxycarbonyl)indoline. A 1-L, two-necked, round-bottomed flask, equipped with a reflux condenser fitted with a calcium chloride-filled drying tube, a

85

pressure-equalizing addition funnel, and a magnetic stirring bar, is charged with 113.5 g (0.52 mol) of di-tert-butyl dicarbonate (Note 1) and 200 mL of tetrahydrofuran. Through the dropping funnel, 59.6 g (0.50 mol) of indoline is added to the reaction mixture with stirring over 30 min which maintains a steady evolution of carbon dioxide (Note 2). The reaction mixture is stirred at room temperature for an additional 3 hr followed by removal of the solvent with a rotary evaporator. The residual liquid is distilled under reduced pressure to give 107.6-107.8 g of 1-(tert-butoxycarbonyl)-indoline (98% yield) as a colorless oil, bp 83-84°C at 0.1 mm, which on standing solidifies, mp 42-45°C (Note 3).

B. *1-(tert-Butoxycarbonyl)-7-indolinecarboxaldehyde.* An oven-dried, 2-L, three-necked, round-bottomed flask, equipped with an argon inlet, 200-mL pressure-equalizing dropping funnel fitted with a rubber septum, a low temperature thermometer, and a 4.5-cm egg-shaped magnetic stirring bar (Note 4), is flushed with argon and charged with 32.9 g (0.15 mol) of 1-(tert-butoxycarbonyl)indoline, 27.2 mL (0.18 mol) of N,N,N',N'-tetramethylethylenediamine (TMEDA) (Note 5), and 750 mL of anhydrous ether (Note 6). The dropping funnel is charged with 167 mL (0.18 mol) of a 1.08 M solution of sec-butyllithium in cyclohexane via syringe (Note 7). Under a positive pressure of argon, the flask is immersed in a dry ice-acetone bath. When the internal temperature has reached ca. -70°C, a white precipitate of the starting material appears. The solution of sec-butyllithium is added dropwise to this suspension with rapid stirring over 30 min while keeping the internal temperature below -70°C. The dropping funnel is rinsed with 10 mL of anhydrous ether. The light brown mixture is stirred for an additional 2 hr at -78°C, during which time the precipitate completely dissolves and the color of the reaction mixture becomes deep brown. After the 2 hr has elapsed, 17.0 mL (0.22 mol) of N,N-dimethylformamide (DMF) is added dropwise through the addition funnel over a 10-min period (Note 8). After stirring the resulting mixture for 30 min at -78°C, 50 mL of saturated aqueous ammonium chloride solution is added through the dropping funnel over 15 min, and the cooling bath is removed (Note 9). When the internal temperature of the reaction mixture has reached -50°C, 50 mL of water is added dropwise, whereupon the deep orange color becomes light yellow. The reaction mixture is allowed to warm to 0°C and then poured into a 2-L separatory funnel containing water (200 mL). After thorough mixing, the layers are separated, and the aqueous phase is extracted with two 200-mL portions of ether. The combined organic layers are washed twice with 200 mL of saturated aqueous sodium chloride solution, dried over sodium sulfate, filtered, and concentrated under reduced pressure on a rotary evaporator followed by exposure to oil pump vacuum to remove as much of the volatiles as possible. The residue is chromatographed on 750 g of silica gel (Note 10), using a carefully packed 8-cm diameter glass column, and a mixture of toluene and ethyl acetate (30 : 1) as the eluent (Note 11). The appropriate fractions are combined and concentrated to give crude 1-(tert-butoxycarbonyl)-7-indolinecarboxaldehyde (23.6-25.5 g, 64-69% yield) as a light yellow solid. Recrystallization of this solid from a mixture of ethyl acetate and hexane affords the pure aldehyde (20.5-21.1 g, 55-57% yield in two crops) as practically colorless needles, mp 86.5-87.5°C (Note 12).

C. *7-Indolinecarboxaldehyde*. A 300-mL Erlenmeyer flask fitted with a magnetic stirring bar is charged with 150 mL of concentrated (36%) hydrochloric acid. To this magnetically stirred solution, 19.8 g (0.08 mol) of finely powdered 1-(tert-butoxycarbonyl)-7-indolinecarboxaldehyde is added portionwise over 15 min at room temperature. The starting material gradually dissolves accompanied by gas evolution. After 2 hr, the resulting orange-colored solution is poured into a 2-L beaker containing crushed ice (500 g). To this mixture, 120 mL of 28% aqueous ammonia solution is added slowly with stirring, and the resulting mixture containing a yellow precipitate is transferred to a 2-L separatory funnel and extracted with four 100-mL portions of dichloromethane. The combined extracts are washed with 100 mL of saturated aqueous sodium chloride solution, dried over sodium sulfate, filtered, and concentrated on a rotary evaporator. The resulting yellowish brown oil is passed through 300 g of silica gel (Note 10) packed in an 8-cm diameter glass column using a mixture of toluene and ethyl acetate (30:1) as an eluent. The yellow eluates are combined, and the solvent is removed to give pure 7-indolinecarboxaldehyde (10.6-11.2 g, 90-95% yield) as a yellow solid, mp 48.5-49°C (Note 13).

2. Notes

1. Di-tert-butyl dicarbonate was purchased from Wako Pure Chemical Industries, Ltd. and used without further purification.

2. Indoline was purchased from Tokyo Kasei Kogyo Co., Ltd. and distilled under reduced pressure before use.

3. The spectral data for 1-(tert-butoxycarbonyl)indoline are as follows: IR (KBr) cm^{-1}: 1700 (C=O); ^{1}H NMR (400 MHz, CDCl$_3$) δ: 1.56 (s, 9 H), 3.07 (t, 2 H, J = 8.5), 3.96 (t, 2 H, J = 8.5), 6.91 (dt, 1 H, J = 7.5, 1.0), 7.11-7.17 (m, 2 H), 7.4-8.0 (extremely broad, 1 H); ^{13}C NMR (100 MHz, CDCl$_3$) δ: 27.3, 28.5, 47.6, 80.8, 114.7, 122.1, 124.7, 127.4, 131.1, 142.8, 152.7.

4. For efficient stirring, a powerful magnetic stirrer should be used. The submitters employed Super Stirrer Model MS-2 manufactured by Ishii Laboratory Works Co., Ltd.

5. N,N,N',N'-Tetramethylethylenediamine was distilled from powdered calcium hydride and stored under argon.

6. Ether was distilled from sodium benzophenone ketyl under nitrogen.

7. sec-Butyllithium was purchased from Kanto Chemical Co., Inc. and used after titration with 2,5-dimethoxybenzyl alcohol.[2]

8. N,N-Dimethylformamide was distilled from powdered calcium hydride and stored over 3 Å molecular sieves under argon.

9. When the reaction mixture was warmed to room temperature (4 hr - stirring after removal of cooling bath) before quenching with aqueous ammonium chloride solution, 7-indolinecarboxaldehyde (9.8-10.1 g, 44-46%) was obtained directly after silica gel column chromatography (SiO$_2$, 750 g, toluene elution).[3] However, the product was contaminated by small amounts of impurities, and attempted purification by recrystallization (ether-pentane) caused considerable loss of the main product.

10. Merck silica gel 60 (230-400 mesh) (No. 9385) was used.

11. Compound 1, a major by-product of this reaction, was eluted after 1-(tert-butoxycarbonyl)-7-indolinecarboxaldehyde on chromatography (ca. 10% yield). This compound could be formed by condensation of the C-7 lithiated 1-(tert-butoxycarbonyl)indoline with the non-lithiated starting material. Compound 1 has mp 192-195°C (decomp.) after recrystallization from ethyl acetate.

The submitters attempted inverse addition of 1-(tert-butoxycarbonyl)indoline to the ether solution of sec-BuLi-TMEDA complex at -78°C in order to suppress this side reaction. Formation of 1 was certainly decreased, but C-2 lithiation was observed as the other side reaction.

12. Spectral data of 1-(tert-butoxycarbonyl)-7-indolinecarboxaldehyde are as follows: IR (KBr) cm^{-1}: 1700, 1675 (C=O); ^{1}H NMR (400 MHz, CDCl$_3$) δ: 1.51 (s, 9 H), 3.07 (t, 2 H, J = 8.0), 4.17 (t, 2 H, J = 8.0), 7.10 (t, 1 H, J = 7.5), 7.36 (dq, 1 H, J = 7.5, 1.0), 7.64 (dd, 1 H, J = 7.5, 1.0), 10.11 (s, 1 H); ^{13}C NMR (100 MHz, CDCl$_3$) δ: 28.2, 49.9, 82.5, 124.0, 125.0, 126.1, 129.2, 134.5, 143.7, 153.9, 189.6.

13. Spectral data of 7-indolinecarboxaldehyde are as follows: IR (KBr) cm^{-1}: 1650 (C=O); ^{1}H NMR (400 MHz, CDCl$_3$) δ: 3.05 (t, 2 H, J = 8.5), 3.78 (t, 2 H, J = 8.5), 6.60 (dd, 1 H, J = 8.0, 7.0), 7.17 (dq, 1 H, J = 7.0, 1.0), 7.27 (dd, 1 H, J = 8.0, 1.0), 9.82 (s, 1 H) (NH absorption not observed); ^{13}C-NMR (100 MHz, CDCl$_3$) δ: 27.8, 47.1, 116.0, 116.1, 129.2, 130.7, 131.1, 153.5, 192.4.

Waste Disposal Information

All toxic materials were disposed of in accordance with "Prudent Practices for Disposal of Chemicals from Laboratories"; National Academy Press; Washington, DC, 1983.

3. Discussion

The procedure described here offers a general route to 7-substituted indolines.[3] The method is based on the directed ortho-lithiation of N-(tert-butoxycarbonyl)aniline derivatives.[4] The tert-butoxycarbonyl group seems to be essential for C-7 selective lithiation, since other directing groups so far reported promote C-2 metalation on the indoline ring.[5] The C-7 selective lithiation of 1-(tert-butoxycarbonyl)indoline is in contrast to the C-2 selective lithiation of 1-(tert-butoxycarbonyl)indole.[6]

The C-7 lithio species reacts successfully with a wide range of electrophiles (chlorotrimethylsilane, tributyltin chloride, diphenyl disulfide, iodine, 1,2-

dibromoethane, hexachloroethane, iodomethane, carbon dioxide, DMF, aromatic and aliphatic aldehydes).[3] Lithiation occurs selectively at C-7 even in the presence of moderately ortho-directing methoxy or chloro groups on the aromatic ring.[3] The tert-butoxycarbonyl group is a well-established protective group for amine functionality and can be easily removed under a variety of reaction conditions.[7] Since indolines are readily oxidized to indoles,[8] this method should be useful for the preparation of 7-substituted indoles, which are not readily prepared by using conventional methodologies.[9]

Two other methods for the C-7 selective functionalization of indoline have been reported. Somei developed C-7 selective thallation of 1-acetylindoline and applied it to the synthesis of 7-substituted indoles.[10] Lo reported a synthesis of 7-benzoylindoline[11] by using Sugasawa's boron trichloride-mediated ortho-acylation of aniline derivatives.[12] The present method is superior to these procedures since a greater diversity of functionality can be introduced, the metalation exhibits high regioselectivity, and the use of highly toxic reagents such as thallium tris(trifluoroacetate) can be avoided.

1. Department of Chemistry, Nagasaki University, 1-14 Bunkyo-machi, Nagasaki 852, Japan.

2. Winkle, M. R.; Lansinger, J. M.; Ronald, R. C. *J. Chem. Soc., Chem. Commun.* **1980**, 87.

3. Iwao, M.; Kuraishi, T. *Heterocycles* **1992**, *34*, 1031.

4. (a) Muchowski, J. M.; Venuti, M. C. *J. Org. Chem.* **1980**, *45*, 4798; (b) Stanetty, P.; Koller, H.; Mihovilovic, M. *J. Org. Chem.* **1992**, *57*, 6833; (c) Beak, P.; Lee, W.-K. *Tetrahedron Lett.* **1989**, *30*, 1197; (d) Beak, P.; Lee, W. K. *J. Org. Chem.* **1993**, *58*, 1109.

W.-K. *Tetrahedron Lett.* **1989**, *30*, 1197; (d) Beak, P.; Lee, W. K. *J. Org. Chem.* **1993**, *58*, 1109.

5. (a) Meyers, A. I.; Hellring, S. *Tetrahedron Lett.* **1981**, *22*, 5119; (b) Katritzky, A. R.; Sengupta, S. *J. Chem. Soc., Perkin Trans. I* **1989**, 17; (c) Meyers, A. I.; Milot, G. *J. Org. Chem.* **1993**, *58*, 6538.

6. Hasan, I.; Marinelli, E. R.; Lin, L.-C. C.; Fowler, F. W.; Levy, A. B. *J. Org. Chem.* **1981**, *46*, 157.

7. Greene, T. W.; Wuts, P. G. M. "Protective Groups in Organic Synthesis", 2nd ed.; John Wiley & Sons: New York, 1991; p. 327.

8. (a) Inada, A.; Nakamura, Y.; Morita, Y; *Chem. Lett.* **1980**, 1287; (b) Ketcha, D. M. *Tetrahedron Lett.* **1988**, *29*, 2151.

9. For recent approaches to the preparation of 7-substituted indoles, see: (a) Moyer, M. P.; Shiurba, J. F.; Rapoport, H. *J. Org. Chem.* **1986**, *51,* 5106; (b) Bartoli, G.; Palmieri, G.; Bosco, M.; Dalpozzo, R. *Tetrahedron Lett.* **1989**, *30*, 2129; (c) Dobson, D.; Todd, A.; Gilmore, J. *Synth. Comm.* **1991**, *21*, 611; (d) Dobson, D. R; Gilmore, J.; Long, D. A. *Synlett* **1992**, 79.

10. (a) Somei, M.; Saida, Y. *Heterocycles* **1985**, *23*, 3113; (b) Somei, M.; Saida, Y.; Funamoto, T.; Ohta, T. *Chem. Pharm. Bull.* **1987**, *35*, 3146; (c) Somei, M.; Funamoto, T.; Ohta, T. *Heterocycles* **1987**, *26*, 1783; (d) Somei, M.; Kawasaki, T.; Ohta, T. *Heterocycles* **1988**, *27*, 2363.

11. Lo, Y. S.; Walsh, D. A.; Welstead, Jr., W. J.; Mays, R. P.; Rose, E. K.; Causey, D. H.; Duncan, R. L. *J. Heterocycl. Chem.* **1980**, *17*, 1663.

12. Sugasawa, T.; Toyoda, T.; Adachi, M.; Sasakura, K. *J. Am. Chem. Soc.* **1978**, *100*, 4842.

Appendix

Chemical Abstracts Nomenclature (Collective Index Number); (Registry Number)

1-(tert-Butoxycarbonyl)indoline: 1H-Indole-1-carboxylic acid, 2,3-dihydro-1,1-dimethylethyl ester (13); (143262-10-6)

Di-tert-butyl dicarbonate: Formic acid, oxydi-, di-tert-butyl ester (8); Dicarbonic acid, bis(1,1-dimethylethyl)ester (9); (24424-99-5)

Indoline (8); 1H-Indole, 2,3-dihydro- (9); (496-15-1)

N,N,N',N'-Tetramethylethylenediamine: Ethylenediamine, N,N,N',N'-tetramethyl- (8); 1,2-Ethanediamine, N,N,N',N'-tetramethyl- (9); (110-18-9)

sec-Butyllithium: Lithium, sec-butyl- (8); Lithium, (1-methylpropyl)- (9); (598-30-1)

N,N-Dimethylformamide: CANCER SUSPECT AGENT: Formamide, N,N-dimethyl- (8,9); (68-12-2)

REGIO- AND STEREOSELECTIVE INTRAMOLECULAR HYDROSILYLATION OF α-HYDROXY ENOL ETHERS: 2,3-syn-2-METHOXYMETHOXY-1,3-NONANEDIOL

(1,3-Nonanediol, 2-(methoxymethoxy)-, (R*,R*)-(±)-)

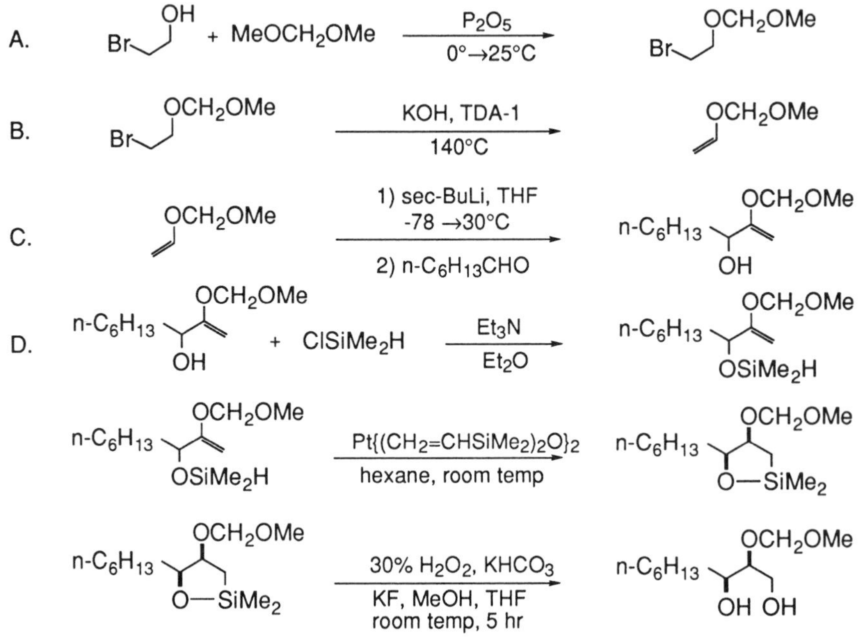

A.
OH
Br
+ MeOCH2OMe
P2O5
0°→25°C
OCH2OMe
Br
B.
OCH2OMe
Br
KOH, TDA-1
140°C
OCH2OMe
C.
OCH2OMe
1) sec-BuLi, THF
-78 →30°C
2) n-C6H13CHO
n-C6H13
OCH2OMe
OH
D.
n-C6H13
OCH2OMe
OH
+ ClSiMe2H
Et3N
Et2O
n-C6H13
OCH2OMe
OSiMe2H
Pt{(CH2=CHSiMe2)2O}2
hexane, room temp
O—SiMe2
30% H2O2, KHCO3
KF, MeOH, THF
room temp, 5 hr
OH OH

Submitted by Kohei Tamao,[1a] Yoshiki Nakagawa,[1b] Yoshihiko Ito.[1b]

Checked by Michael R. Reeder, Lisa M. Reeder, and Robert K. Boeckman, Jr.

1. Procedure

A. 2-Bromoethyl methoxymethyl ether. A 1-L, single-necked, round-bottomed flask is equipped with a magnetic stirring bar and a gas inlet valve (Note 1). The flask is purged with argon, and charged with 125.0 g of 2-bromoethanol (1.00 mol), and 500 mL of dimethoxymethane (Notes 2 and 3). The resulting mixture is cooled to 0°C in an ice-water bath. As a single portion, 71.0 g (0.5 mol) of solid phorphorus pentoxide (P_2O_5) is added with stirring (Note 4). The resulting mixture, which becomes quite viscous because of the insoluble P_2O_5, is stirred at 0°C for 10 min, then warmed to room temperature and stirred for an additional 7 hr during which time the viscosity of the mixture decreases. The mixture is transferred to a 2-L separatory funnel, 200 mL of water is added, and the resulting mixture is extracted three times with 50 mL of ether. The combined organic extracts are washed successively with 100 mL of water and 100 mL of saturated aqueous sodium carbonate solution, dried over magnesium sulfate, filtered, and carefully concentrated with a rotary evaporator under reduced pressure (20 - 30 mm) at room temperature. The residue is distilled under reduced pressure to afford 131.3 g (78%) of 2-bromoethyl methoxymethyl ether as a colorless liquid, bp 74-78°C (40 mm) (Note 5).

B. *Methoxymethyl vinyl ether.* A 200-mL, two-necked, round-bottomed flask is equipped with a stopper, magnetic stirring bar (Note 6), and an efficient reflux condenser that is attached through a thick rubber tube to a clean trap, cooled in a dry ice-acetone bath, whose outlet is equipped with a calcium chloride drying tube. The flask is charged with 74.2 g (440 mmol) of 2-bromoethyl methoxymethyl ether, 58 g (880 mmol) of 85% potassium hydroxide pellets (Note 7), and 7.09 g (2.2 mmol) of tris[2-(2-methoxyethoxy)ethyl]amine (TDA-1) (Note 8). The mixture is heated to 140°C with stirring. The clear liquid phase gradually becomes cloudy and begins to reflux. After heating for 27 hr, consumption of the starting material is confirmed by GLC (Note 9). The calcium chloride tube is replaced with an additional cold trap that is attached to a water aspirator. The reaction mixture is cooled to 110°C and a vacuum is applied using the water aspirator for 2 hr, during which time volatile material is collected in the first cold trap. After the flask is cooled to room temperature, the first trap is disconnected and allowed to warm to room temperature. The liquid is transferred to a

50-mL, round-bottomed flask. A small amount of water (about 1 mL) is usually present and is removed with a syringe. The organic layer is distilled at atmospheric pressure into an ice-cooled receiver to afford 26.6 - 28.2 g (69-72%) of methoxymethyl vinyl ether as a colorless liquid, bp 64-70°C (Note 10). The product is dried and kept over 4Å molecular sieves.

C. 2-Methoxymethoxy-1-nonen-3-ol. A 500-mL, two-necked, round-bottomed flask is equipped with a rubber septum, a magnetic stirring bar, and a pressure-equalizing dropping funnel bearing a three-way stopcock connected to a mineral oil bubbler. The flask is evacuated, purged with nitrogen, and charged with 200 mL of dry tetrahydrofuran (THF) (Note11) and 8.83 g of methoxymethyl vinyl ether (100 mmol). The solution is cooled to -78°C with a dry ice-acetone bath, and then 88.7 mL (100 mmol) of sec-butyllithium (1.13 M in cyclohexane) is added dropwise over 15 min with stirring via syringe. After the addition is complete, the yellow mixture is allowed to warm to -30°C over 30 min and stir for 1 hr. The mixture is recooled to -78°C and 11.44 g (100 mmol) of heptanal is added dropwise over 5 min through the dropping funnel. The reaction mixture is allowed to warm to room temperature over 30 min and to stir for 3 hr. A 100-mL portion of saturated aqueous sodium chloride solution is added, the mixture is transferred to a 1-L separatory funnel, mixed thoroughly, and the organic layer separated. The aqueous layer is extracted three times with 50 mL of diethyl ether. The combined organic phases are dried over sodium sulfate (Na_2SO_4), filtered, and concentrated under reduced pressure with a rotary evaporator. The residue is distilled under reduced pressure to afford 12.6-13.1 g (62-65%) of 2-methoxymethoxy-1-nonen-3-ol as a colorless liquid, bp 82-84°C (0.4 mm) (Note 12).

D. 2-Methoxymethoxy-1,3-nonanediol. A 500-mL, single-necked, round-bottomed flask, equipped with a three-way stopcock connected to a bubbler as above and a magnetic stirring bar, is charged with 14.79 g (73.1 mmol) of 2-methoxymethoxy-1-nonen-3-ol (Note 6). The flask is evacuated, then purged with nitrogen, and charged with 350 mL of dry diethyl ether and 8.88 g (87.7 mmol) of triethylamine (Note 13). After the mixture is cooled to 0°C in an ice-water bath, 8.30 g (87.7 mmol) of chlorodimethylsilane is added slowly with stirring. A large quantity of salt appears immediately. The white suspension is stirred at room temperature for 1 hr. The resulting mixture is then filtered through a sintered-glass Büchner funnel and the filter cake is washed thoroughly with dry hexane (Note 14). The combined filtrate and washings are concentrated under reduced pressure with a rotary evaporator. Some salts usually remain in the residue and are removed by dilution of the residue with an additional 50 mL of dry hexane followed by filtration, washing the filter cake with hexane, and concentration of the combined filtrate and washings as above. This procedure for removal of salts is repeated two to three times until the colorless residual liquid remains clear after concentration (Note 15). The residual liquid is finally subjected to high vacuum (0.4 mm) for 2 hr to complete removal of volatile material, and then transferred to a 300-mL, round-bottomed flask, equipped with a three-way stopcock and a magnetic stirring bar. The flask is evacuated, purged with nitrogen, and charged with 73 mL of dry hexane. To this mixture is added 1.21 mL (0.35 mmol) of the previously prepared platinum catalyst solution (Note 16) at room temperature with stirring. An exothermic reaction ensues almost immediately and the temperature of the mixture rises to about 30°C in 15 min, during which time the color of the clear mixture changes to yellow. The exothermic reaction ceases in about 1 hr. After the reaction mixture is stirred for an additional 1.5 hr at room temperature, completion of the reaction is confirmed by analysis by ^{1}H NMR (Note 17). A 7.3-g portion of activated carbon powder is then added to the lightly colored mixture to effect removal of organoplatinum species (Note 18). The black suspension is stirred for about 12 hr, and suction filtered through Celite into a 500-mL, two-necked, round-bottomed flask. The filter cake is washed with dry hexane, and the combined filtrate and washings are concentrated by rotary evaporation (Note 19). The flask containing the residue is equipped with a magnetic stirring bar and a thermometer, and is kept open to the air throughout subsequent manipulations. The flask is charged with 73 mL of THF, 73 mL of methanol (Note 20), 7.32 g (73.1 mmol) of solid potassium hydrogen carbonate, and 8.5 g (146 mmol) of solid potassium fluoride (Note 21). To the stirred mixture is added 29.7 mL (263 mmol) of 30% hydrogen peroxide slowly at room temperature (Note 22). A somewhat cloudy upper organic layer and a milky white lower aqueous layer result. After several minutes, an exothermic reaction begins which is controlled by intermittent, brief cooling with a water bath to maintain the temperature at about 40°C. The exothermic reaction ceases in about 2 hr. The mixture is stirred at room temperature for an additional 3 hr (Note 23). The excess hydrogen peroxide is decomposed by careful dropwise addition of a 50% aqueous sodium thiosulfate pentahydrate solution (~20 mL) with stirring over 20 min during which time the temperature is maintained near 30°C by intermittent cooling with an ice bath (Note 24). A negative starch-iodide test is observed after addition is complete (Note 25). The mixture is concentrated by rotary evaporator at 40-50°C under reduced pressure (water aspirator). To the residual aqueous suspension is added 150 mL of anhydrous ethanol and the mixture is stirred vigorously until the sticky, pale yellow precipitate becomes granular. The mixture is concentrated as above to give a pale yellow solid. The ethanol addition and evaporation procedure is repeated once more to insure thorough removal of water. The residue is then diluted with 150 mL of diethyl ether, the mixture is filtered with suction, and the filter cake is washed well with two 100-mL portions of diethyl ether. The combined filtrate and washings are concentrated by rotary evaporation. The residual, crude product is purified by chromatography on 250 g of silica gel with hexane/ethyl acetate (1:1 v/v) as eluent to give 9.34 - 10.62 g (58-66%) of 2-methoxymethoxy-1,3-nonanediol (R_f = 0.2) as a colorless liquid (Note 26). The syn and anti isomers cannot be separated by column chromatography. However, separation can be effected by GLC which showed the isomer ratio to be 98:2 syn/anti (Note 27).

2. Notes

1. A powerful magnetic stirrer and a large rugby ballshaped stirring bar can be used. The checkers suggest use of an overhead mechanical stirrer because of the viscosity of the mixture in the initial stages of the reaction, which makes efficient magnetic stirring difficult.

2. 2-Bromoethanol was obtained from Aldrich Chemical Company, Inc., and was distilled prior to use.

3. Dimethoxymethane was purchased from the Aldrich Chemical Company, Inc., and used as received.

4. Reagent grade phosphorus pentoxide was obtained from the J. T. Baker Chemical Company and used as received.

5. 2-Bromoethyl methoxymethyl ether exhibits the following spectral properties: ^{1}H NMR (200 MHz, CDCl$_3$) δ: 3.38 (s, 3 H), 3.49 (t, 2 H, J = 6.2), 3.85 (t, 2 H, J = 6.2), 4.66 (s, 2H); ^{13}C NMR (50 MHz, CDCl$_3$) δ: 30.74, 55.45, 67.79, 96.54; IR (neat) cm^{-1}: 2952, 2900, 1148, 1118, 1072, 1036; Anal. Calcd for C$_4$H$_9$BrO$_2$: C, 28.43; H, 5.37. Found: C, 28.39; H, 5.36.

6. A powerful magnetic stirrer and a large rugby ballshaped stirring bar can be used. The checkers suggest use of an overhead mechanical stirrer because of the large quantity of salts produced in the reaction, which makes efficient magnetic stirring difficult.

7. Potassium hydroxide pellets are better than finely ground powder, because the powder readily solidifies into a mass upon heating causing the reaction to slow down. Sonication does not facilitate the reaction.

8. TDA-1 is not essential, but accelerates the reaction considerably.

9. Isothermal GLC analysis was performed on a 30-m x 0.32-mm ID HP-5 capillary column containing 5% phenylmethylsilicone as a stationary phase at 50°C (retention time of the bromo ether is 3.8 min).

10. Methoxymethyl vinyl ether exhibits the following spectral properties: ^{1}H NMR (200 MHz, C_6D_6) δ: 3.11 (s, 3 H), 4.09 (dd, 1 H, J = 6.6, 1.4), 4.54 (s, 2 H), 4.61 (dd, 1 H, J = 14.2, 1.4), 6.32 (dd, 1 H, J = 14.2, 6.6); ^{13}C NMR (50 MHz, C_6D_6) δ: 55.37, 90.71, 95.28, 149.94.

11. Tetrahydrofuran was distilled from sodium benzophenone ketyl.

12. Since 2-methoxymethoxy-1-nonen-3-ol tends to decompose on attempted GLC analysis, the purity is best checked by ^{1}H NMR. 2-Methoxymethoxy-1-nonen-3-ol exhibits the following spectral properties: ^{1}H NMR (200 MHz, C_6D_6) δ: 0.91 (t (br), 3 H, J = 7), 1.20-1.95 (m, 11 H), 3.17 (s, 3 H), 4.02 (q (br), 1 H, J = 5), 4.35 (s, 2 H), 4.75 (d, 1H, J = 6), 4.78 (d, 1 H, J = 6); ^{13}C NMR (50 MHz, C_6D_6) δ: 14.27, 22.99, 25.88, 29.64, 32.20, 35.73, 55.71, 72.95, 84.21, 93.87, 162.63; IR (neat) cm[-1]: 3448, 2940, 2864, 1644, 1156, 1096, 1020; Anal. Calcd for $C_{11}H_{22}O_3$: C, 65.31; H, 10.96. Found: C, 65.48; H, 11.17.

13. Triethylamine was distilled from calcium hydride. Freshly distilled triethylamine should be used, because any impurities in aged triethylamine may poison the catalyst in the next step.

14. Hexane was distilled from sodium.

15. Since the O-silylated product is somewhat sensitive to moisture, the filtration process should be performed quickly.

16. Platinum catalyst preparation: a mixture of 1 g of chloroplatinic acid hexahydrate, 12.4 mL (22 equiv) of 1,3-divinyltetramethyldisiloxane (an excess), and 1.2 mL of water is heated to 65-70°C for 3.5 hr with stirring.[2] The initial mixture becomes homogeneous after ~30 min. The solution is then cooled to room

temperature, and 5 mL of aqueous saturated sodium bicarbonate ($NaHCO_3$) solution is added. The upper, yellow-orange, organic layer was separated, dried over Na_2SO_4, and filtered affording a solution (~0.29 M) of the platinum catalyst in disiloxane that was stored in the refrigerator until use.

17. An aliquot is taken into an NMR tube, solvent is removed under reduced pressure, C_6D_6 is added and the sample is analyzed by ^{1}H-NMR. The following spectral and analytical data for the cyclic siloxanes were obtained. The syn/anti ratio (95/5) was determined by capillary GLC (Note 9) except that a temperature program was employed [100°C for 2 min, then increased 15°C/min to 250°C (retention times 6.46 min (anti), 6.66 min (syn)]. Each isomer was isolated by preparative GLC: Syn: ^{1}H NMR (200 MHz, C_6D_6) δ: 0.17 (s, 3 H), 0.29 (s, 3 H), 0.78 (dd, 1 H, J = 15, 5.5), 0.92 (t (br), 3 H, J = 7), 1.03 (dd, 1 H, J = 15, 5), 1.20-1.65 (m, 7 H), 1.65-2.00 (m, 3 H), 3.23 (s, 3 H), 3.89 (dt (br), 1 H, J = 9, 4), 4.12 (q (br), 1 H, J = 5), 4.48 (d, 1 H, J = 7), 4.65 (d, 1 H, J = 7); ^{13}C NMR (50 MHz, C_6D_6) δ: 0.28, 1.14, 14.30, 18.39, 23.04, 26.72, 29.96, 32.18, 32.33, 55.28, 77.91, 80.91, 95.16; IR (neat) cm^{-1}: 2963, 2864, 1470, 1254, 1150, 1102, 1046, 872. Anti: ^{1}H NMR (200 MHz, C_6D_6) δ: 0.15 (s, 3 H), 0.23 (s, 3 H), 0.83 (dd, 1 H, J = 14.2, 7.8), 0.91 (t (br), 3 H, J = 6), 1.18 (dd, 1 H, J = 14, 6), 1.25-1.85 (m, 10 H), 3.26 (s, 3 H), 3.90-4.14 (m, 2 H), 4.57 (d, 1 H, J = 7), 4.73 (d, 1 H, J = 7). Anal. Calcd for $C_{13}H_{28}O_3Si$ (isomeric mixture): C, 59.95, H, 10.84. Found: C, 59.67; H, 11.02.

18. It is essential to remove all organoplatinum species since these species rapidly decompose hydrogen peroxide in the next oxidation step.

19. The filtrate is ordinarily colorless but may possibly be brown, owing to the presence of trace amounts of organoplatinum species.

20. Commercial reagent grade THF and methanol are used without further purification.

21.   Reagent grade anhydrous potassium fluoride was purchased from J. T. Baker Chemical Company. This material must be weighed quickly because it is highly hygroscopic.

22.   Slow, not vigorous, evolution of small bubbles of oxygen due to a trace amount of residual organoplatinum species may be observed.

23.   Since the hydrosilylation products decompose on silica gel, consumption of the starting material cannot be monitored by TLC. Only the formation of the product diol can be detected (R_f = 0.2, hexane/ethyl acetate 1:1).

24.   *Caution: The thiosulfate solution **MUST NOT** be added in one portion; since a sudden, violent, uncontrollable reaction could occur.*

25.   If the starch-iodide test is still positive for the presence of peroxide, additional thiosulfate solution should be added until a negative starch-iodide test is obtained.

26.   The major syn isomer of 2-methoxymethoxy-1,3-nonanediol exhibits the following spectral properties: ^{1}H NMR (200 MHz, $CDCl_3$) δ: 0.86 (t (br), 3 H, J = 7), 1.18-1.50 (m, 10 H), 2.52 (d, 1 H, J = 5), 2.95 (dd, 1 H, J = 8, 4), 3.35-3.50 (m, 4 H, including a sharp singlet at 3.44), 3.55-3.82 (m, 3 H), 4.71 (d, 1 H, J = 7) 4.80 (d, 1 H, J = 7); ^{13}C NMR (50 MHz, $CDCl_3$) δ: 14.02, 22.54, 25.43, 29.23, 31.72, 33.31, 55.89, 63.22, 71.80, 84.12, 97.43; IR (neat) cm^{-1}: 3428, 2936, 1470, 1154, 1108, 1032. Anal. Calcd for $C_{11}H_{24}O_4$ (isomer mixture): C, 59.97; H, 10.98. Found: C, 59.78; H, 11.04.

27.   The syn/anti ratio (98:2) was determined by capillary GLC on a 30-m x 0.32-mm ID HP-5 capillary column containing 5% phenylmethylsilicone as the stationary phase using the following temperature program: 100°C for 2 min, then increase 15°C/min to 250°C (retention times 6.96 min (anti), 7.55 min (syn)). The submitters report that the syn/anti ratio can be determined after conversion to the acetonide by treatment with excess dimethoxypropane and catalytic acid. By this procedure, the

isomer ratio was found to be 94.4:5.6 (syn/anti). The submitters also report that the isomers can be separated easily by chromatography on silica gel: R_f = 0.31 for the syn isomer and R_f = 0.45 for the anti isomer (hexane/ethyl acetate 1:1). The acetonides exhibit the following spectral properties: syn: ^{1}H NMR (200 MHz, CDCl$_3$) δ: 0.85 (t (br), 3 H, J = 7), 1.15-1.75 (m, 16 H, including two sharp singlets at 1.41 and 1.42), 3.33 (q, 1 H, J = 2.0), 3.39 (s, 3 H), 3.80-4.04 (m, 3 H), 4.64 (d, 1 H, J = 7.1), 4.78 (d, 1 H, J = 7.1); ^{13}C NMR (50 MHz, CDCl$_3$) δ: 14.01, 18.89, 22.53, 24.92, 29.17, 31.05, 31.72, 55.71, 62.82, 70.00, 71.30, 95.29, 98.52 (Only five signals are observed for the hexyl group). Anal. Calcd for $C_{14}H_{28}O_4$: C, 64.58; H, 10.84. Found: C, 64.42; H, 11.11. Anti: ^{1}H NMR (200 MHz, CDCl$_3$) δ: 0.81 (t (br), 3 H, J = 6.5), 1.10-1.80 (m, 16 H, including two sharp singlets at 1.30 and 1.38), 3.23-3.38 (m, 4 H, including a sharp singlet at 3.29), 3.50-3.68 (m, 2 H), 3.90 (dd, 1 H, J = 11.4, 5.2), 4.54 (d, 1 H, J = 6.8), 4.59 (d, 1 H, J = 6.8); ^{13}C NMR (50 MHz, CDCl$_3$) δ: 14.08, 19.76, 22.62, 25.06, 28.23, 29.24, 31.83, 32.34, 55.62, 63.39, 72.11, 74.48, 96.49, 98.60. The checkers found the determination of the syn/anti ratio by this procedure to be unreliable because of the presence of interfering impurities formed during conversion to the acetonide.

Waste Disposal Information

All toxic materials were disposed of in accordance with "Prudent Practices for Disposal of Chemicals from Laboratories"; National Academy Press; Washington, DC, 1983.

3.   Discussion

This procedure for the preparation of 2-methoxymethoxy-1,3-nonanediol is a modification of that reported by the submitters.[3]   Among a wide variety of methodologies for stereoselective construction of polyoxygenated skeletons, the most

103

straightforward is the stereoselective introduction of two oxygen functionalities onto carbon-carbon double bonds, as represented by the Sharpless epoxidation[4] or osmium tetroxide oxidation of allyl alcohols.[5] An alternative route may be envisioned by anti-Markovnikov hydration of enol ether derivatives, as in the present method, but such an approach has so far been rarely studied. The only other reported method is hydroboration which affords syn/anti ratios in the range of 83/17 to 3/97 in essentially the same systems as those examined in this procedure.[6] The method described here is a highly stereoselective route to 2,3-syn isomers of 1,2,3-triol skeletons from α-hydroxy enol ethers.

This procedure consists of the synthesis of a precursor, methoxymethyl vinyl ether, an α-hydroxy enol ether, and the intramolecular hydrosilylation of the latter followed by oxidative cleavage of the silicon-carbon bonds. The first step, methoxymethylation of 2-bromoethanol, is based on Fujita's method.[7] The second and third steps are modifications of results reported by McDougal and his co-workers.[8] Dehydrobromination of 2-bromoethyl methoxymethyl ether to methoxymethyl vinyl ether[8] was achieved most efficiently with potassium hydroxide pellets[8,9] rather than with potassium tert-butoxide as originally reported for dehydrobromination of the tetrahydropyranyl analog.[10] Potassium tert-butoxide was effective for the dehydrobromination, but formed an adduct of tert-butyl alcohol with the vinyl ether as a by-product in substantial amounts. Methoxymethyl vinyl ether is lithiated efficiently with sec-butyllithium in THF[8] and, somewhat less efficiently, with n-butyllithium in tetrahydrofuran. Since lithiation of simple vinyl ethers such as ethyl vinyl ether requires tert-butyllithium,[11] metalation may be assisted by the methoxymethoxy group in the present case.

In the final step for this large scale preparation, the hydroxyl group is silylated with chlorodimethylsilane and triethylamine to effect introduction of the dimethylsilyl group. Alternatively, (diethylamino)dimethylsilane (HMe$_2$SiNEt$_2$) or 1,1,3,3-tetra methyldisilazane (HMe$_2$Si)$_2$NH in the presence of a catalytic amount of ammonium chloride as a promoter can also be used. In these cases, the filtration step can be omitted since removal of the generated amine and the excess silylamine affords the crude oily product that can be directly employed in the hydrosilylation step.[3,12] Intra-molecular hydrosilylation of α-hydroxy enol ethers proceeds readily with a neutral platinum catalyst such as the vinylsiloxane/platinum(0) complex[2] since the most commonly used catalyst, chloroplatinic acid hydrate (H$_2$PtCl$_6$•6H$_2$O), is not suitable for use with acid-sensitive enol ethers. Similar regio- and stereoselective intramolecular hydrosilylations can also be achieved with rhodium catalysts.[13] For removal of organoplatinum species prior to the oxidation, activated carbon seems to be more efficient than the disodium salt of ethylenediaminetetraacetic acid (EDTA•2Na) originally used.[3] Oxidative cleavage of the silicon-carbon bond is performed under standard conditions.[14]

Finally, the regio- and stereoselectivities of the intramolecular hydrosilylation of allyl alcohols, homoallyl alcohols, and allylamines are summarized in Scheme 1. While these results have been obtained with the dimethylsilyl (HMe$_2$Si) group in the presence of a platinum catalyst, stereoselectivity, but not regioselectivity, can be controlled by other silyl groups such as the diphenyl (HPh$_2$Si) or di(isopropyl)silyl [H(i-Pr)$_2$Si] groups.

Scheme 1

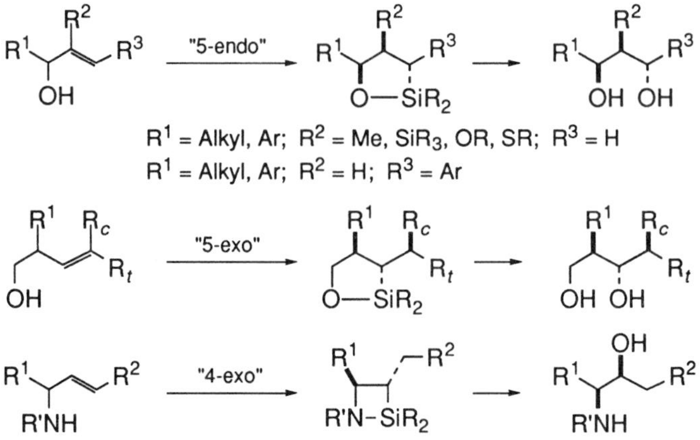

R^1 = Alkyl, Ar; R^2 = Me, SiR$_3$, OR, SR; R^3 = H
R^1 = Alkyl, Ar; R^2 = H; R^3 = Ar

In the allylic alcohol series having a terminal olefin, the reaction proceeds in the 5-endo fashion to give five-membered ring compounds selectively with the syn stereoisomer being predominant, regardless of the nature of the catalyst, platinum or rhodium.[15,16] Stereoselectivity increases with an increase in bulk of the allylic substituent. In addition, much higher selectivities are attained with enol ethers (R^2 = OMOM, OTHP, SR)[3] in comparison to other systems [(R^2 = Me, SiMe$_2$(O-i-Pr)[17]]. Stereoselectivities in cyclic enol ethers appear to be sensitive to the reaction conditions.[18] It has also been shown that the 5-endo cyclization proceeds smoothly with allyl alcohols with a terminal aryl group (R^3 = Ar), but not with an alkyl group.[19] In homoallyl alcohols, the 5-exo type of intramolecular hydrosilylation proceeds to form five-membered cyclic compounds and eventually 1,3-diols by oxidative cleavage. Two types of 1,2-stereoselection are possible, one being anti-controlled by the allylic substituent and, the other depending upon the olefin geometries, anti to the cis substituent R$_c$ and syn to the trans R$_t$.[15] Catalytic, asymmetric, intramolecular hydrosilylation is also possible with rhodium catalysts in the presence of optically active phosphine ligands.[13,19] In contrast, allylamines form four-membered cyclic products via 4-exo ring closure in the platinum-catalyzed hydrosilylation from which syn-1,2-amino alcohols are obtained highly stereoselectively after oxidative cleavage.[20]

1.   (a) Institute for Chemical Research, Kyoto University, Uji, Kyoto 611, Japan; (b) Department of Synthetic Chemistry, Faculty of Engineering, Kyoto University, Kyoto 606, Japan.

2.   (a) Chandra, G.; Lo, P. Y.; Hitchcock, P. B.; Lappert, M. F. *Organometallics* **1987**, *6*, 191; (b) Hitchcock, P. B.; Lappert, M. F.; Warhurst, N. J. W. *Angew. Chem., Int. Ed. Engl.* **1991**, *30*, 438, and references cited therein.

3.   Tamao, K.; Nakagawa, Y.; Arai, H.; Higuchi, N.; Ito, Y. *J. Am. Chem. Soc.* **1988**, *110*, 3712.

4.   Johnson, R. A.; Sharpless, K. B. In *Comprehensive Organic Synthesis*; Trost, B. M.; Fleming, I., Ed.; Pergamon Press: Oxford, U.K., 1991; Vol. 7, pp. 389-436.

5.   Haines, A. H. In *Comprehensive Organic Synthesis*; Trost, B. M.; Fleming, I., Ed.; Pergamon Press: Oxford, U.K., 1991; Vol. 7, pp. 437-448.

6.   McGarvey, G. J.; Bajwa, J. S. *Tetrahedron Lett.* **1985**, *26*, 6297 and references cited therein.

7.   Fuji, K.; Nakano, S.; Fujita, E. *Synthesis* **1975**, 276.

8.   McDougal, P. G.; Rico, J. G.; VanDerveer, D. *J. Org. Chem.* **1986**, *51*, 4492.

9.   Boeckman, R. K., Jr.; Bruza, K. J. *Tetrahedron* **1981**, *37*, 3997.

10. d'Angelo, J. *Bull. Soc. Chim. Fr.* **1969**, 181.

11. Baldwin, J. E.; Höfle, G. A.; Lever, O. W., Jr. *J. Am. Chem. Soc.* **1974**, *96*, 7125.

12. Tamao, K. *Yuki Gosei Kagaku Kyokaishi* **1988**, *46*, 861; *Chem. Abstr.* **1989**, *111*, 153857.

13. Tamao, K.; Tohma, T.; Inui, N.; Nakayama, O.; Ito, Y. *Tetrahedron Lett.* **1990**, *31*, 7333.

14. Tamao, K.; Ishida, N.; Ito, Y., Kumada, M. *Org. Synth.* **1990**, *69*, 96.

15. (a) Tamao, K.; Nakajima, T.; Sumiya, R.; Arai, H.; Higuchi, N.; Ito, Y. *J. Am. Chem. Soc.* **1986**, *108*, 6090; (b) Tamao, K.; Tanaka, T.; Nakajima, T.; Sumiya, R.; Arai, H.; Ito, Y. *Tetrahedron Lett.* **1986**, *27*, 3377.

16. Anwar, S.; Davis, A. P. *Proc. R. Ir. Acad.* **1989**, *89B*, 71.

17. Tamao, K.; Egawa, Y.; Nakagawa, Y.; Ito, Y., unpublished results (1990).

18. Curtis, N. R.; Holmes, A. B. *Tetrahedron Lett.* **1992**, *33*, 675.

19. Bergens, S. H.; Noheda, P.; Whelan, J.; Bosnich, B. *J. Am. Chem. Soc.* **1992** , *114*, 2121.

20. Tamao, K.; Nakagawa, Y.; Ito, Y. *J. Org. Chem.* **1990**, *55*, 3438.

Appendix
Chemical Abstracts Nomenclature (Collective Index Number); (Registry Number)

2,3-syn-2-Methoxymethoxy-1,3-nonanediol: 1,3-Nonanediol, 2-(methoxymethoxy)-, (R*,R*)-(±)- (12); (114675-32-0)

2-Bromoethyl methoxymethyl ether: Ethane, 1-bromo-2-(methoxymethoxy)- (12); (112496-94-3)

2-Bromoethanol: HIGHLY TOXIC: Ethanol, 2-bromo- (8,9); (540-51-2)

Dimethoxymethane: Methane, dimethoxy- (8,9); (109-87-5)

Phosphorus pentoxide: Phosphorus oxide (8,9); (1314-56-3)

Methoxymethyl vinyl ether: Ethene, (methoxymethoxy)- (10); (63975-05-3)

Tris[2-(2-methoxyethoxy)ethyl]amine: TDA-1: Ethanamine, 2-(2-methoxyethoxy)-N,N-bis[2-(2-methoxyethoxy)ethyl}- (10); (70384-51-9)

2-Methoxymethoxy-1-nonen-3-ol: 1-Nonen-3-ol, 2-(methoxymethoxy)-, (±)- (12); (114675-31-9) sec-Butyllithium: Lithium, sec-butyl- (8); Lithium, (1-methylpropyl)- (9); (598-30-1)

Heptanal (8,9); (111-71-7)

Triethylamine (8); Ethanamine, N,N-diethyl- (9); (121-44-8)

Chlorodimethylsilane: Silane, chlorodimethyl- (8,9); (1066-35-9)

Chloroplatinic acid hexahydrate: Platinate(2-), hexachloro-, dihydrogen (8); Platinate(2-), hexachloro-, dihydrogen (OC-6-11)- (9); (16941-12-1)

1,3-Divinyltetramethyldisiloxane: Disiloxane, 1,1,3,3-tetramethyl-1,3-divinyl- (8); Disiloxane, 1,3-diethenyl-1,1,3,3-tetramethyl- (9); (2627-95-4)

Hydrogen peroxide (8,9); (7722-84-1)

A GENERAL PROCEDURE FOR MITSUNOBU INVERSION OF STERICALLY HINDERED ALCOHOLS: INVERSION OF MENTHOL.
(1S,2S,5R)-5-METHYL-2-(1-METHYLETHYL)CYCLOHEXYL 4-NITROBENZOATE
(Cyclohexanol, 5-methyl-2-(1-methylethyl)-, 4-nitrobenzoate, [1S-(1α,2α,5β)]-)

Submitted by Jeffrey A. Dodge, Jeffrey S. Nissen, and Misti Presnell.[1]
Checked by Masami Okabe, Ruen Chu Sun, and David Coffen.

1. Procedure

(1S,2S,5R)-5-Methyl-2-(1-methylethyl)cyclohexyl 4-nitrobenzoate. A 250-mL, three-necked, round-bottomed flask is equipped with a stirring bar, nitrogen inlet, rubber septum, and thermometer. The flask is charged with 3.00 g of (1R,2S,5R)-(-)-menthol (19.2 mmol), 12.9 g of 4-nitrobenzoic acid (77.2 mmol), 20.1 g of triphenylphosphine (PPh_3) (76.6 mmol) (Note 1), and 150 mL of tetrahydrofuran (Note 2). The flask is immersed in an ice bath, and 12.1 mL of diethyl azodicarboxylate (77 mmol) is added dropwise at a rate such that the temperature of the reaction mixture is maintained below 10°C (Note 3). Upon completion of the addition (Note 4), the flask is removed from the ice bath and the solution is allowed to stir at room temperature overnight (14 hr) and subsequently at 40°C for 3 hr (Note 5). The reaction mixture is

110

cooled to room temperature, diluted with 150 mL of ether, and washed twice with 100 mL portions of saturated aqueous sodium bicarbonate solution. The aqueous layers are combined and back-extracted with 100 mL of ether. The combined organic layers are dried over sodium sulfate. Excess solvent and other volatile reaction components are completely removed under reduced pressure initially on a rotary evaporator and then under high vacuum (approximately 0.2 mm for 3 hr at 30°C) (Note 6). The resulting semi-solid is suspended in 40 mL of ether and the suspension is allowed to stand at room temperature overnight (Note 7). The mixture is stirred while 20 mL of hexanes is slowly added (Note 8). The resulting white solid is filtered under vacuum and the filter cake is washed with 200 mL of 50% (v/v) ether-hexanes. The solvent is removed from the filtrate on a rotary evaporator under reduced pressure to give a yellow oil that is dissolved in 10 mL of methylene chloride (Note 9) and diluted with 40 mL of 8% ether-hexanes. The solution is applied to a flash chromatography column (Note 10) and eluted with 8% ether-hexanes to give 5.03 g (85.6%) of pure nitrobenzoate ester as a white crystalline solid (Note 11).

2. Notes

1. (1R,2S,5R)-(-)-Menthol, 4-nitrobenzoic acid, triphenylphosphine, and diethyl azodicarboxylate were purchased from Aldrich Chemical Company, Inc., and used without further purification.

2. Anhydrous tetrahydrofuran (THF) was purchased from Aldrich Chemical Company, Inc. (Sure/Seal™ bottle). The checkers used a freshly opened bottle of certified grade THF from Fisher Scientific Company.

3. Since diethyl azodicarboxylate decomposes when warmed, the reaction temperature is maintained at < 10°C during the addition of this reagent at which time a slight exothermic reaction occurs.

4. 4-Nitrobenzoic acid is not entirely soluble in the tetrahydrofuran solution, resulting in a heterogeneous mixture. Upon addition of diethyl azodicarboxylate the reaction becomes homogeneous within several minutes and turns yellow-orange.

5. Shorter reaction times (2-5 hr) result in slightly decreased yields (65-75%) of inverted product. Lower yields (73-75%) were realized by the checkers when stirring at 40°C was omitted.

6. Removal of all residual tetrahydrofuran is critical to the success of the subsequent precipitation. The submitters recommend high vacuum removal of the solvent. Use of "house vacuum" (4-10 mm) also proved effective if applied to the crude sample for several hours.

7. Precipitation of the undesired reaction by-products (reduced diethyl azodicarboxylate, triphenylphosphine oxide) occurs after the product mixture is suspended in ether.

8. Extractive workup can be deleted, in which case sonication is required at this stage. Sonication appears to initiate further crystallization of by-products, as well as minimizing the amount of oil formed during the addition of hexanes.

9. The yellow oil is not sufficiently soluble in the chromatographic eluant (8% ether in hexanes) to provide efficient loading of the sample on the column. Thus, the crude material is initially dissolved in a small amount of dichloromethane.

10. Chromatography was performed via the method of Still[2] using an 80-mm i.d. column of 300 g of silica gel (230-400 mesh).

11. The physical properties are as follows: mp 93-95°C; ^{1}H NMR (300 MHz, CDCl$_3$) δ: 0.87-1.23 (m, 12 H), 1.44-1.56 (m, 2 H), 1.64-1.70 (m, 1 H), 1.71-1.90 (m, 2 H), 2.05-2.13 (m, 1 H), 5.50 (s, 1 H), 8.20 (d, 2 H, J = 8.5), 8.29 (d, 2 H, J = 8.9); ^{13}C NMR (75 MHz, CDCl$_3$,) δ: 20.0, 20.8, 22.0, 25.3, 26.7, 29.0, 29.3, 34.6, 46.8, 73.0, 123.4, 130.5, 136.3, 150.3, 163.8; IR (CDCl$_3$) cm^{-1}: 1726, 1540, 1293.

Waste Disposal Information

All toxic materials were disposed of in accordance with "Prudent Practices for Disposal of Chemicals from Laboratories"; National Academic Press; Washington, DC, 1983.

3. Discussion

Since its introduction in 1967,[3] the Mitsunobu reaction has been widely used in the refunctionalization of alcohols, in particular, for inversion. However, Mitsunobu inversions of hindered alcohols have been problematic resulting in low yields or unused starting material. A simple modification of the standard Mitsunobu procedure using 4-nitrobenzoic acid (instead of benzoic or acetic acid) results in significantly improved yields of inverted product for sterically hindered alcohols.[4] The procedure outlined here provides experimental details for the inversion of menthol, a representative, hindered, secondary alcohol, using 4-nitrobenzoic acid as the acidic coupling partner in the Mitsunobu process. The experimental procedure is a modification of that reported by Martin and Dodge.[4] Improvements in the solvent (tetrahydrofuran rather than benzene) and a practical method for the removal of undesired by-products (e.g., $EtCO_2NHNHCO_2Et$ and triphenylphosphine oxide) have led to optimization of reaction conditions.

Previously documented methods for menthol inversion under standard Mitsunobu conditions (benzoic acid, PPh_3, diethyl azodicarboxylate) result in low yields[4] (27%). More effective methods have been reported using extended reaction periods in refluxing toluene via a formic acid/N,N'-dicyclohexylcarbodiimide-mediated transformation[5] (20-92 hr, 80%). For hindered alcohols in general, representative methods for inverting alcohol stereochemistry necessitate conversion of the alcohol to

113

a leaving group such as a mesylate or triflate, followed by S_N2 displacement with a carboxylate nucleophile (typically cesium acetate or propionate).[6] Other prevalent methods include (a) oxidation followed by stereoselective reduction and (b) intramolecular delivery (for suitably functionalized substrates) of the requisite oxygen functionality. One inherent advantage of the Mitsunobu reaction is compatibility with a diverse array of functional groups, due in large part to the mild, essentially neutral reaction conditions.

1. Lilly Research Laboratories, Eli Lilly and Company, Indianapolis, IN 46285.

2. Still, W. C.; Kahn, M.; Mitra, A. *J. Org. Chem.* **1978**, *43*, 2923.

3. Mitsunobu, O.; Yamada, M. *Bull. Chem. Soc. Jpn.* **1967**, *40*, 2380; Mitsunobu, O. *Synthesis* **1981**, 1-28.

4. Martin, S. F.; Dodge, J. A. *Tetrahedron Lett.* **1991**, *32*, 3017.

5. Kaulen, J. *Angew. Chem., Int. Ed. Engl.* **1987**, *26*, 773; Kaulen, J. Ger. Offen. DE 3 511 210, 1986; *Chem. Abstr.* **1987**, *106*, 32087g.

6. For example, see Torisawa, Y.; Okabe, H.; Ikegami, S. *Chem. Lett.* **1984**, 1555.

Appendix
Chemical Abstracts Nomenclature (Collective Index Number);
(Registry Number)

(1R,2S,5R)-Menthol: Menthol, (-)- (8); Cyclohexanol, 5-methyl-2-(1-methylethyl)-, [1R-(1α,2β,5α)]- (9); (2216-51-5)

(1S,2S,5R)-1-(4-Nitrobenzoyl)-2-(1-methylethyl)-5-methylcyclohexan-1-ol: Cyclohexanol, 5-methyl-2-(1-methylethyl)-, 4-nitrobenzoate, [1S-(1α,2α,5β)]- (9); (27374-00-1)

4-Nitrobenzoic acid: Benzoic acid, p-nitro- (8); Benzoic acid, 4-nitro- (9); (62-23-7)

Triphenylphosphine: Phosphine, triphenyl- (8,9); (603-35-0)

Diethyl azodicarboxylate: Formic acid, azodi-, diethyl ester (8); Diazenedicarboxylic acid, diethyl ester (9); (1972-28-7)

A SIMPLE AND CONVENIENT METHOD FOR THE OXIDATION OF ORGANOBORANES USING SODIUM PERBORATE: (+)-ISOPINOCAMPHEOL

(Bicyclo[3.1.1]heptan-3-ol, 2,6,6-trimethyl-, [1S-(1α,2β,3α,5α)]-)

A.

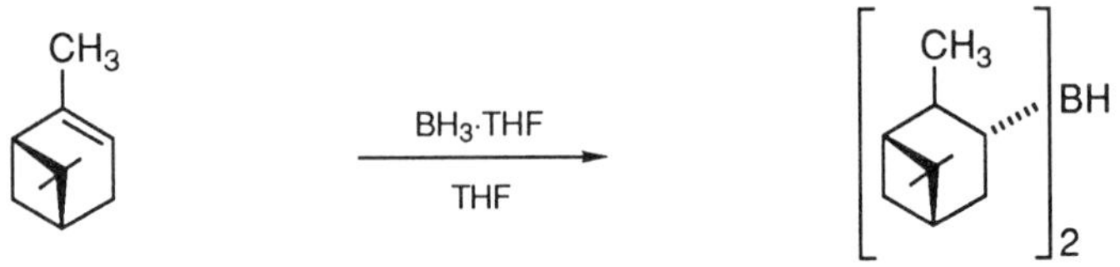
CH3
BH3·THF
THF
CH3
BH
2

B.

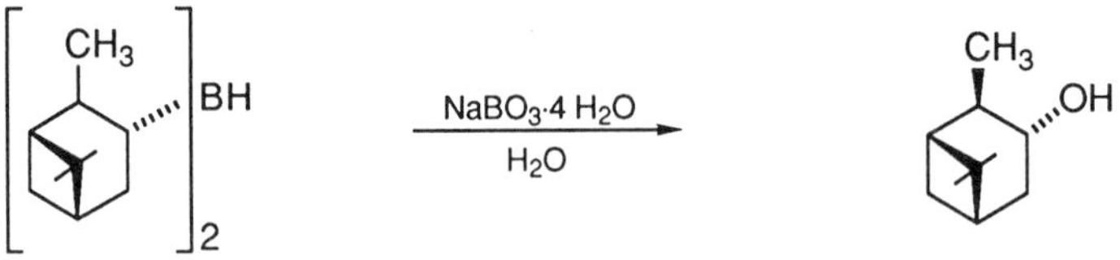
CH3
BH
2
NaBO3·4 H2O
H2O
CH3
OH

Submitted by George W. Kabalka, John T. Maddox, Timothy Shoup, and Karla R. Bowers.[1]
Checked by C. Huart and Leon Ghosez.

1. Procedure

Caution! This procedure should be conducted in an efficient fume hood to assure the adequate removal of hydrogen, a flammable gas which forms explosive mixtures with air.

A. (+)-Diisopinocampheylborane. A dry, 250-mL, three-necked, round-bottomed flask, equipped for magnetic stirring, and with a nitrogen inlet vented through a mercury bubbler, a rubber septum, and a thermometer, is flushed with nitrogen and charged with 13.75 g (0.101 mol) of (-)-α-pinene and 25 mL of tetrahydrofuran (Notes 1-3). The mixture is cooled to 0°C in an ice-water bath, magnetic stirring is initiated, and 58.0 mL (0.055 mol) of 0.95 M borane-tetrahydrofuran solution (Note 4) is added via syringe at a rate such that the temperature of the reaction mixture remains below 5°C. After the addition is complete, the cooling bath is removed. The reaction mixture is allowed to warm to room temperature and stir for 2 hr to ensure complete reaction.[2]

B. *(+)-Isopinocampheol.* To the stirred solution of (+)-diisopinocampheyl-borane in tetrahydrofuran, prepared above, 50 mL of distilled water is *slowly* added dropwise via syringe [*CAUTION!*] (Note 5) followed by the slow addition of 16.41 g (0.107 mol) of solid sodium perborate tetrahydrate (Note 6) through an appropriate addition funnel at a rate such that the temperature of the reaction mixture does not exceed 35°C (Note 7). Stirring is continued at room temperature (22°C) for 2 hr to ensure completion of the oxidation reaction. The contents of the flask are then poured into 70 mL of ice-cold water in a separatory funnel. After thorough mixing, the organic layer is removed and the aqueous layer is extracted twice with 25 mL of ether. The combined ether extracts are washed twice with 20-mL portions of water and then with 50 mL of saturated aqueous sodium chloride solution. The ether layer is dried over anhydrous magnesium sulfate, filtered, and concentrated on a rotary evaporator. The crude product is purified by short-path vacuum distillation to give 14.1-14.3 g (91-92%) of isopinocampheol, bp 68°C (0.7 mm), as white needles (mp 53-55°C) that crystallize in the receiving flask (Note 8). Recrystallization from pentane gives pure isopinocampheol as needles, mp 54-55°C (uncorrec.), $[\alpha]_D^{25}$ 34.4° (benzene, *c* 10) indicating 96.4% enantiomeric purity based on the highest reported literature value of 35.7° (Note 9).[4]

2. Notes

1. All glassware was predried at 140°C for at least 4 hr, assembled hot, flame dried, and cooled under a stream of nitrogen.

2. (1S)-(-)-α-Pinene [99%, 98% optical purity, $[\alpha]_D^{25}$ -50.6° neat] was purchased from Aldrich Chemical Company, Inc., and distilled under reduced pressure from lithium aluminum hydride before use.[2]

3. Tetrahydrofuran was distilled under nitrogen from sodium benzophenone ketyl.

4. Borane-tetrahydrofuran complex (1.0 M) was obtained from Aldrich Chemical Company, Inc., and the concentration of the solution determined using the literature procedure.[3]

5. Since the hydroboration only proceeds to the dialkylborane stage, a large amount of hydrogen is evolved on hydrolysis. Very slow dropwise addition of water and adequate ventilation are recommended.

6. Sodium perborate tetrahydrate was purchased from Aldrich Chemical Company, Inc. and used as received.

7. During the addition of sodium perborate the reaction flask is kept in a water bath (25°C).

8. An air condenser is employed for the distillation. The receiving flask is immersed in an ice bath.

9. The product exhibits the following spectral properties: IR (melt) cm^{-1}: 3300 (OH), 2930, 1472, 1450, 1384, 1367, 1050, 1015; ^{1}H NMR (250 MHz, CDCl$_3$) δ: 0.92 (s, 3 H), 1.04 (d, 1 H, J = 9), 1.13 (d, 3 H, J = 7), 1.22 (s, 3 H), 1.67-2.11 (m, 5 H), 2.30-2.58 (m, 2 H), 4.06 (dt, 1H); ^{13}C NMR (62.87 MHz, CDCl$_3$) δ: 20.7, 23.7, 27.7, 34.4, 38.1, 39.1, 41.8, 47.8, 47.9, 71.7.

Waste Disposal Information

All toxic materials were disposed of in accordance with "Prudent Practices for Disposal of Chemicals from Laboratories"; National Academic Press; Washington, DC, 1983.

3. Discussion

This procedure illustrates the simplest and most convenient method for oxidizing organoboranes, in the present example a dialkylborane. It uses sodium perborate,[5] an inexpensive, safe, and easily handled reagent, as the oxidizing agent. The reaction proceeds under mild conditions and the yield of the product alcohol is generally as high or higher than that obtained in the sodium hydroxide/hydrogen peroxide oxidation procedure.[6] Thus, the sodium perborate method is an attractive alternative to the base/hydrogen peroxide oxidation procedure. In the case described above, the perborate procedure produces higher yields of (-)-isopinocampheol than the sodium hydroxide/hydrogen peroxide procedure with comparable stereoselectivity.[7] Although the mechanism of the oxidation has not been investigated in detail, sodium perborate does not appear to be acting as a simple mixture of hydrogen peroxide and sodium borate.[8-11] Presumably, borate is a more effective leaving group (Scheme 1) than hydroxide ion which is generated during oxidation by hydrogen peroxide .

Scheme 1

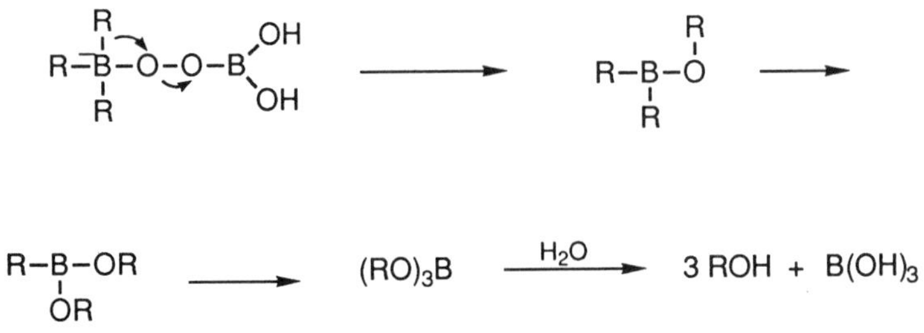

Sodium perborate, owing to its stability, commercial availability, and ease of handling, should prove to be a popular reagent for oxidizing organoboranes. Some representative examples of the oxidation of organoboranes bearing a variety of alkyl and aryl groups are listed in Table I.[12]

TABLE I

SODIUM PERBORATE OXIDATION OF ORGANOBORANES [a]

Alkene	Reagent	Product	Yield(%)[b]
	BH₂	OH [c]	93
	BH	OH [d]	99
	BH₃	OH	92
	BH₃	OH	86
	BH₂		84
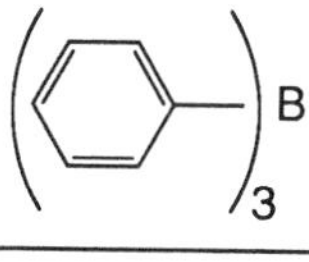 B [e]	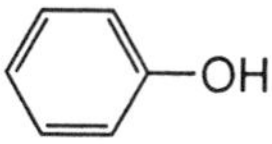 OH	87

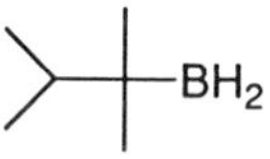

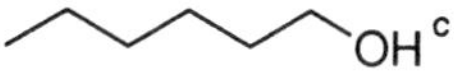

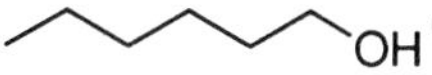

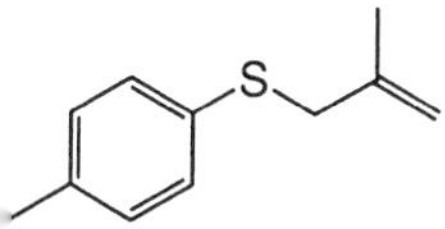

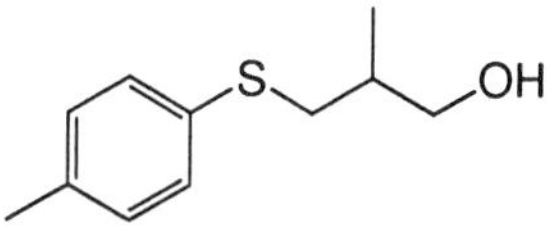

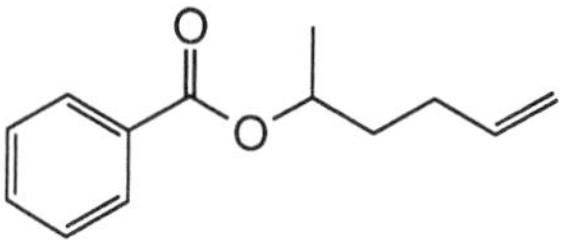

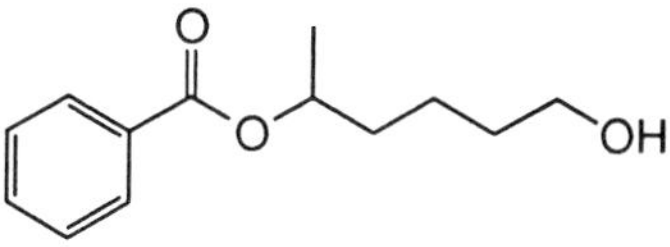

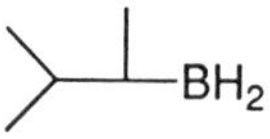

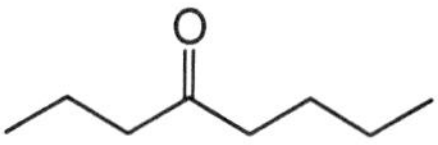

[a] The organoboranes were formed via the hydroboration of the alkene listed. [b] Isolated yield. [c] One equivalent of 2,3-dimethyl-2-butanol was isolated in addition to 2 equiv of 2-hexanol. [d] One equivalent of 1,4-cyclooctanediol was isolated in addition to the 1-hexanol. [e] Triphenyl-borane was purchased from Aldrich Chemical Company, Inc.

121

1.   Department of Chemistry, University of Tennessee, Knoxville, TN 37996-1600.

2.   Brown, H. C.; Singaram, B. *J. Org. Chem.* **1984**, *49*, 945.

3.   Brown, H. C.; Kramer, A. W.; Levy, A. B.; Midland, M. M. "Organic Syntheses via Boranes"; Wiley: New York, 1975; pp. 241-245.

4.   Brown, H. C.; Yoon, N. M. *Isr. J. Chem.* **1977**, *15*, 12.

5.   "Kirk-Othmer Encyclopedia of Chemical Technology", 3rd ed.; Wiley: New York, 1978; Vol. 3, pp. 944-945, Vol. 17, pp. 2-9.

6.   Lane, C. F. *J. Org. Chem.* **1974**, *39*, 1437.

7.   Lane, C. F.; Daniels, J. J. *Org. Synth., Coll. Vol. VI* **1988**, 719.

8.   Dehmlow, E. V.; Vehre, B. *New J. Chem.* **1989**, *13*, 117.

9.   Hansson, A. *Acta Chem. Scand.* **1961**, *15*, 934.

10. McKillop, A.; Tarbin, J. A. *Tetrahedron* **1987**, *43*, 1753.

11. Matteson, D. S.; Moody, R. J. *J. Org. Chem.* **1980**, *45*, 1091.

12. (a) Kabalka, G. W.; Shoup, T. M.; Goudgaon, N. M. *Tetrahedron Lett.* **1989**, *30*, 1483; (b) Kabalka, G. W.; Shoup, T. M.; Goudgaon, N. M. *J. Org. Chem.* **1989**, *54*, 5930.

Appendix

Chemical Abstracts Nomenclature (Collective Index Number); (Registry Number)

(+)-Isopinocampheol: Bicyclo[3.1.1]heptan-3-ol, 2,6,6-trimethyl, [1S-(1α,2β,3α,5α)] (9); (24041-60-9)

(-)-α–Pinene: Bicyclo[3.1.1]hept-2-ene, 2,6,6-trimethyl-, (1S)- (9); (7785-26-4)

Borane-tetrahydrofuran complex: Furan, tetrahydro-, compd. with borane (1:1) (8,9); (14044-65-6)

Sodium perborate tetrahydrate: Perboric acid sodium salt, tetrahydrate (8,9); (10486-00-7)

DOUBLE HYDROXYLATION REACTION FOR CONSTRUCTION OF THE CORTICOID SIDE CHAIN: 16α-METHYLCORTEXOLONE

(Pregn-4-ene-3,20-dione, 17,21-dihydroxy-16-methyl-, (16α)-(±)-)

Submitted by Yoshiaki Horiguchi, Eiichi Nakamura, and Isao Kuwajima.[1]

Checked by Ronald W. Regenye, Miguel Pagan, and David L. Coffen.

1.   Procedure

Caution: Hexamethylphosphoramide (HMPA) has been identified as a carcinogen. Glove protection is required during the handling in Part A. In addition, the column chromatography in Part B using chloroform as the eluent should be conducted in a well-ventilated hood.

A. (Z)-16α-Methyl-20-trimethylsiloxy-4,17(20)-pregnadien-3-one (**2**). (All transfers are conducted under dry nitrogen; reagents are introduced into reaction vessels through rubber septa using a syringe.) An oven-dried, 300-mL, two-necked, round-bottomed flask equipped with a magnetic stirring bar, nitrogen-vacuum inlet, and rubber septum is charged with 6.25 g (20 mmol) of 16-dehydroprogesterone (**1**) and 0.20 g (1.0 mmol) of cuprous bromide-dimethyl sulfide complex (Note 1). After the apparatus is flushed with nitrogen, 100 mL of tetrahydrofuran (THF) and 7.7 mL (44 mmol) of hexamethylphosphoramide are added (Note 1). The resulting clear solution, upon cooling to -78°C, becomes a white slurry to which 5.1 mL (40 mmol) of chlorotrimethylsilane is added dropwise (Note 1). To the resulting yellow solution is added 23.7 mL (22 mmol) of a 0.93 M solution of methylmagnesium bromide in THF (Note 2) over a 30-min period. The resulting yellow slurry is then stirred at ~-55 to -60°C (Note 3) for 12 hr followed by addition of 5.6 mL (40 mmol) of triethylamine dropwise (Note 1). The reaction mixture is then poured into a vigorously stirred mixture of 50 mL of saturated aqueous sodium bicarbonate, 50 g of ice, and 200 mL of hexane. After stirring for 15 min, the mixture is transferred to a 1-L separatory funnel, and the organic phase is separated. The remaining aqueous phase is extracted three times with 50-mL portions of hexane. The combined organic phases are washed successively with 50 mL of water and 50 mL of brine, dried over anhydrous magnesium sulfate, and concentrated under reduced pressure to give 6.84-8.67 g of crude (Z)-16α-methyl-20-trimethylsiloxy-4,17(20)-pregnadien-3-one (**2**) as an amorphous white solid. Analysis of crude **2** by ^{1}H NMR indicates a chemical purity of 90-95% and a geometrical ratio of >95% (Z) (Note 4).

B. 16α-Methylcortexolone (**3**). An oven-dried, 1-L, three-necked, round-bottomed flask, equipped with a magnetic stirring bar, nitrogen-vacuum inlet, 200-mL addition funnel topped with a nitrogen inlet, and a rubber septum, is charged with 7.20 g of the crude (Z)-16α-methyl-20-trimethylsiloxy-4,17(20)-pregnadien-3-one (**2**). The apparatus is flushed with nitrogen and 200 mL of methylene chloride (CH$_2$Cl$_2$) is added (Note 1). Quickly under nitrogen flow, the rubber septum is removed from the flask and 12.8 g (128 mmol) of finely powdered, dry potassium bicarbonate (Note 5) is added to the solution, and the flask is resealed with the rubber septum. The flask is then immersed in an ice bath. With vigorous stirring mixture, 100 mL of a 0.5 M solution (50 mmol) of m-chloroperoxybenzoic acid (MCPBA) in CH$_2$Cl$_2$ is added dropwise via the addition funnel over a 2.5-hr period followed by a few mL of CH$_2$Cl$_2$ to rinse the addition funnel (Note 6). TLC is used to monitor the progress of the reaction (Note 7). After stirring the reaction mixture for an additional 10 min after the addition is complete, the addition funnel, nitrogen-vacuum inlet, and rubber septum are removed and 100 mL of aqueous 0.5 M sodium thiosulfate solution is added to the reaction mixture and vigorous stirring maintained at room temperature for 30 min. The mixture is then transferred to a 1-L separatory funnel, and the organic phase is separated. The aqueous phase is extracted three times with 50 mL of CH$_2$Cl$_2$. The combined organic extracts are concentrated on a rotary evaporator. The residue is dissolved in 100 mL of THF, and the solution is acidified to ~ pH 1 by addition of 10 mL of 1 N hydrochloric acid (HCl) to effect desilylation of the 21-trimethylsilyl ether of 16α-methylcortexolone (6) (Note 8). The homogeneous solution is allowed to stand at room temperature for 30 min and then most of the solvent is removed by rotary evaporation under reduced pressure. The residue is dissolved in 300 mL of CH$_2$Cl$_2$, transferred into a 1-L separatory funnel, and the solution washed with 50 mL of saturated aqueous sodium bicarbonate solution. After separation of the organic phase, the aqueous phase is extracted three times with 50-mL portions of CH$_2$Cl$_2$. The combined organic extracts are washed with 50 mL of brine, dried over anhydrous magnesium sulfate, and concentrated under reduced pressure to give 6.0 g of a white solid. Chromatographic purification on silica gel (300 g) with 30~40% ethyl acetate/chloroform eluent gives 2.92 g (40.5%, 2 steps) of 16α-methylcortexolone (**3**) (Note 9).

2. Notes

1. 16-Dehydroprogesterone (**1**) was purchased from Sigma Chemical Company and used without further purification. Cuprous bromide-dimethyl sulfide complex was prepared according to House's procedure.[2] Hexamethyl-phosphoramide, chlorotrimethylsilane, and triethylamine were purchased from Tokyo Kasei Kogyo Co., Ltd., Japan and distilled from calcium hydride (CaH_2). Tetrahydrofuran (THF) was distilled from sodium-benzophenone ketyl immediately prior to use. Methylene chloride was distilled from phosphorus pentoxide (P_2O_5).

2. A THF solution of methylmagnesium bromide was purchased from Tokyo Kasei Kogyo Co., Ltd., Japan and titrated with sec-butyl alcohol using 1,10-phenanthroline as indicator. Rapid addition might raise the internal temperature and use of excess methylmagnesium bromide would cause undesired methylation of the A-ring enone.

3. The reaction temperature was controlled by an electric cooling system. A higher reaction temperature would cause undesired methylation of the A-ring enone.

4. Crude **2** is free from HMPA. The spectral properties of **2** were as follows: IR (neat) cm^{-1}: 1670, 1610, 1265, 1250, 1230; ^{1}H NMR (200 MHz, CDCl$_3$) δ: 0.19 (s, 9 H), 0.90 (s, 3 H), 0.99 (d, 3 H, J = 7.1), 1.09-2.71 (m including two s at 1.19 and 1.79, 25H), 5.729 (s (br), 1 H); ^{13}C NMR (50 MHz, CDCl$_3$) δ: 1.07, 17.1, 17.3, 20.6, 21.3, 22.1, 32.1, 32.9, 33.5, 34.0, 34.2, 34.3, 35.6, 37.3, 38.7, 44.0, 52.1, 54.1, 123.6, 132.5, 139.9, 171.3, 199.2. The geometry was determined based on observed NOEs from 20-methyl to 16β-H and 16α-methyl.

5. Potassium bicarbonate was purchased from Koso Chemical Co., Ltd., Japan. It was finely powdered and dried under reduced pressure (~ 0.1 mm) at ambient temperature over P_2O_5.

6. m-Chloroperoxybenzoic acid (MCPBA) of 85% purity was purchased from Aldrich Chemical Company, Inc. and purified according to Schwartz's procedure[3] to remove any remaining m-chlorobenzoic acid. Slow addition of MCPBA is required to avoid hydrolysis of the transient, intermediate epoxide **4** by rapid formation of free m-chlorobenzoic acid.

7. Progress of the double hydroxylation reaction can be monitored by TLC analysis. The R_f values of the products with 30% ethyl acetate/hexanes as the eluent are as follows: 0.70 for **2**, 0.59 for **5**, 0.29 for **6**, and 0.18 for **7**. Additional MCPBA may be added until the intermediate hydroxy enol silyl ether **5** has completely reacted.

8. Desilylation of **6** can be monitored by TLC analysis. The R_f values of **3** and **6** are 0.32 and 0.67, respectively, with 50% ethyl acetate/hexanes as the eluent.

9. A portion of this compound is recrystallized from 1:1 ethyl acetate/hexanes to yield white plates with mp 194-197°C (Anal. Calcd for $C_{22}H_{32}O_4$: C, 73.30; H, 8.95. Found: C, 73.24; H, 8.98). The spectral properties were as follows: IR (CDCl$_3$) cm^{-1}: 3650-3100, 1705, 1660, 1615; ^{1}H NMR (400 MHz, CDCl$_3$) δ: 0.79 (s, 3 H), 0.93 (d, 3 H, J = 7.3), 0.97-2.49 (m including s at 1.18), 2.62 (s, 1 H), 2.94-3.14 (m, 1 H), 3.20 (s (br), 1 H), 4.30 (dd, 1 H, J = 20, 4.8), 4.62 (dd, 1 H, J = 20, 4.8), 5.73 (s (br), 1 H): ^{13}C NMR (100 MHz, CDCl$_3$) δ: 14.8, 15.2, 17.4, 20.5, 30.4, 32.0, 32.5, 32.8, 33.9, 35.6, 35.7, 36.8, 38.6, 49.7, 49.8, 53.3, 67.8, 90.5, 123.8, 170.8, 199.3, 212.4. The stereochemistry of the 16- and 17-positions were determined based on the observed NOEs from the 18-methyl (δ 0.79, s) to both 16β-H (δ 2.94-3.14, m) and 21-H (δ 4.634, dd). The submitters obtained an overall yield of 68%

127

Waste Disposal Information

All toxic materials were disposed of in accordance with "Prudent Practices for Disposal of Chemicals from Laboratories"; National Academic Press; Washington, DC, 1983.

3. Discussion

The present procedure is an efficient two-step preparation of the 17-dihydroxyacetone side chain with a 16α-methyl substituent from the 16-dehydro-17-acetyl substructure.[4] The D-ring substructure of the product is of pharmaceutical importance as seen in synthetic corticoids such as betamethasone.[5] The two-step conversion consists of 1) conjugate addition of a methyl group into the 16-position and 2) a novel, double hydroxylation of the resultant enol silyl ether.

Although the chlorotrimethylsilane-accelerated conjugate addition of the catalytic methylcopper reagent[6] proceeds at the sterically less congested D-ring enone in a highly chemoselective manner under the reaction conditions discussed in the procedure, a higher reaction temperature and/or use of excess methylmagnesium bromide might cause undesired methylation of the A-ring enone.

Since Hassner's initial report in 1975,[7] oxidation of an enol silyl ether with peracid has been a reliable method for the preparation of α-siloxy and α-hydroxy ketones. However, the submitters have found that, if the enol silyl ether possesses certain structural features, the reaction, with more than two equivalents of the oxidant, affords α,α'-dihydroxylated ketones (i.e., introduction of two oxygen atoms in a single-step) instead of the expected monohydroxylated compounds.[8]

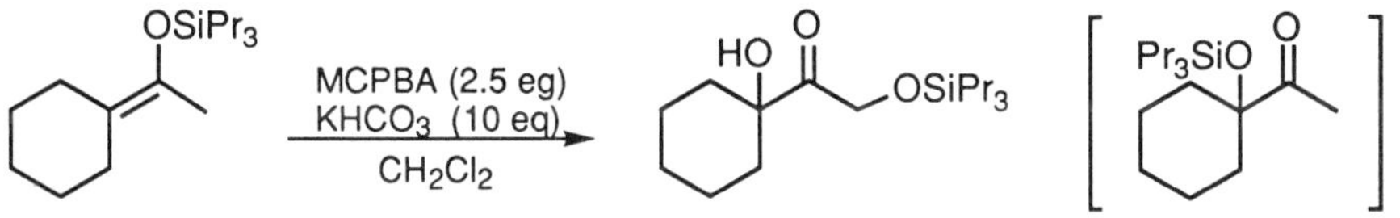

Mechanistic investigations carried out in some depth suggested an interesting reaction pathway (*path a*, Scheme I), in which rearrangement of the intermediate epoxide **B** to the hydroxy enol silyl ether **D** (with loss of H$^{\cdot}$) represents the crucial step. In the normal Hassner reaction (*path b*), rearrangement of epoxide **B** to the siloxy ketone **C** proceeds through migration of the silyl group from the enol oxygen to the epoxide oxygen. The inertness of **C** under the reaction conditions indicated that *path a* and *path b* are independent reactions. The hydroxy enol silyl ether **D** has been shown to be the primary product of the reaction by its isolation upon use of only one equivalent of the oxidant, and its subsequent conversion to **E** upon addition of another equivalent of the oxidant.

Scheme 1

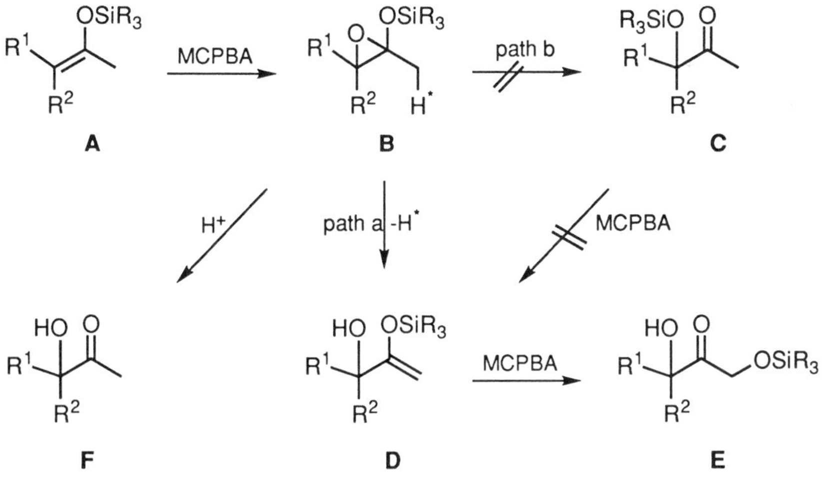

129

The major by-product in the double hydroxylation reaction is the α-hydroxy ketone **F** which forms presumably by protiodesilylation of the transient, intermediate epoxide **B**. In order to exclude free m-chlorobenzoic acid that might cause this side reaction, MCPBA is purified and added very slowly to the substrate in the presence of excess, finely powdered potassium bicarbonate. In the case of the example presented above, the mechanism presumably is as follows:

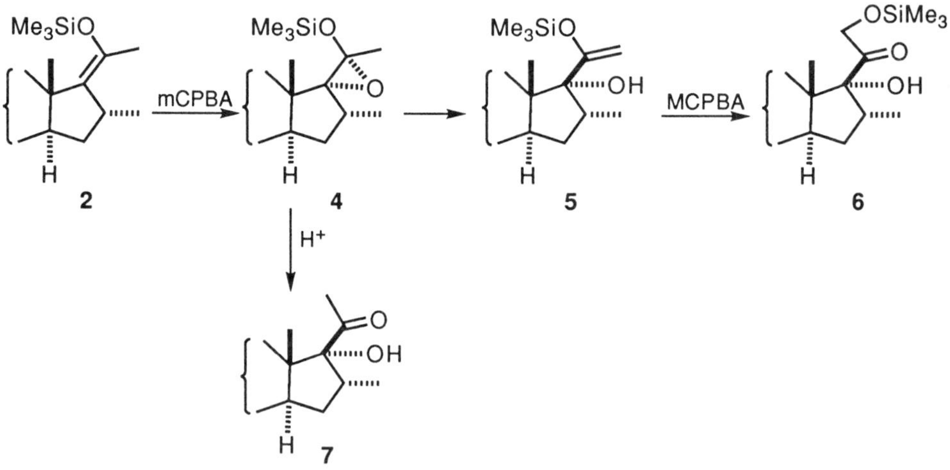

Examples of the double hydroxylation reaction observed for several representative substrates illustrate the scope of this reaction (Table). *Path a* is generally preferred by the internal olefinic isomer of the enol silyl ether of methyl alkyl ketones (entries 1-4, and 9) among which methyl sec-alkyl ketones (entries 1-3, and 9) overwhelmingly prefer the *path a*. Choice of the silyl group substantially affects *path a* vs. *path b* ratio: *path a* becomes the favored pathway when the bulky tripropylsilyl group was used in place of the trimethylsilyl group (cf. entries 4 and 5). Thus steric hindrance at the site of the initial oxidation, the nature of the site of the proton removal (i.e., H* in **B**), and the steric effect of the silyl group all contribute to the relative amounts of the two pathways.

DOUBLE HYDROXYLATION OF ENOL SILYL ETHERS

Entry	Substrate	MCPBA, equiv	Path a: Path b	Combined % Yield	Major product
1	(cyclohexylidene, OSiPr$_3$)	2.5	100:0	72	(HO, O, OSiPr$_3$)
2	(methylcyclohexylidene, OSiMe$_3$)	2.5	100:0	79[a]	(HO, O, OH)
3	C$_8$H$_{17}$ (OSiPr$_3$)	3.3	100:0	74[a]	C$_8$H$_{17}$ (HO, O, OH)
4	(steroid, Pr$_3$SiO)	1.0	100:0	91	(steroid, Pr$_3$SiO, OH)
5	C$_7$H$_{15}$ (OSiPr$_3$)	1.0	63:35	75	C$_7$H$_{15}$ (OSiPr$_3$, OH)
6	C$_7$H$_{15}$ (OSiMe$_3$)	1.0	32:68	94	C$_7$H$_{15}$ (O, OSiMe$_3$)
7	(cyclohexylidene, OSiPr$_3$, C$_5$H$_{11}$)	2.0	25:75	88	(Pr$_3$SiO, O, C$_5$H$_{11}$)
8	(cyclohexene, OSi-t-BuMe$_2$)	2.0	0:100	nd[b]	(t-BuMe$_2$SiO, O)
9	(methylcyclohexene, OSiMe$_3$)	2.0	0:100	nd[b]	(Me$_3$SiO, O)

[a] Isolated after acidic workup. [b] Not determined. A major portion of the initial monooxygenation product was lost by further oxidation with excess MCPBA.

1. Department of Chemistry, Tokyo Institute of Technology, O-okayama, Meguro-ku, Tokyo 152, Japan.

2. House, H. O.; Chu, C.-Y.; Wilkins, J. M.; Umen, M. J. *J. Org. Chem.* **1975**, *40*, 1460.

3. Schwartz, N. N.; Blumbergs, J. H. *J. Org. Chem.* **1964**, *29*, 1976.

4. Horiguchi, Y.; Nakamura, E.; Kuwajima, I. *J. Am. Chem. Soc.* **1989**, *111*, 6257.

5. Taub, D.; Hoffsommer, R. D.; Slates, H. L.; Kuo, C. H.; Wendler, N. L. *J. Am. Chem. Soc.* **1960**, *82*, 4012.

6. Horiguchi, Y.; Matsuzawa, S.; Nakamura, E.; Kuwajima, I. *Tetrahedron Lett.* **1986**, *27*, 4025; Matsuzawa, S.; Horiguchi, Y.; Nakamura, E.; Kuwajima, I. *Tetrahedron* **1989**, *45*, 349.

7. Hassner, A.; Reuss, R. H; Pinnick, H. W. *J. Org. Chem.* **1975**, *40*, 3427.

8. Horiguchi, Y.; Nakamura, E.; Kuwajima, I. *Tetrahedron Lett.* **1989**, *30*, 3323.

Appendix

Chemical Abstracts Nomenclature (Collective Index Number); (Registry Number)

16α-Methylcortexolone: Pregn-4-ene-3,20-dione, 17,21-dihydroxy-16-methyl-, (16α)-(±)- (12); (122405-63-4)

Hexamethylphosphoramide: Phosphoric triamide, hexamethyl- (8,9); (680-31-9)

(Z)-16α-Methyl-20-trimethylsiloxy-4,17(20)-pregnadien-3-one: Pregna-4,17(20)-dien-3-one, 16-methyl-20-[(trimethylsilyl)oxy]-, (16α,17Z)-(±)- (12); (122315-01-9)

16-Dehydroprogesterone: Pregna-4,16-diene-3,20-dione (8,9); (1096-38-4)

Cuprous bromide-dimethyl sufide complex: Copper, bromo[thiobis[methane]]- (9); (54678-23-8)

Chlorotrimethylsilane: Silane, chlorotrimethyl- (8,9); (75-77-4)

Methylmagnesium bromide: Magnesium, bromomethyl- (9); (75-16-1)

Triethylamine (8); Ethanamine, N,N-diethyl- (9); (121-44-8)

m-Chloroperoxybenzoic acid: Peroxybenzoic acid, m-chloro- (8); Benzocarboperoxoic acid, 3-chloro- (9); (937-14-4)

DETRIFLUOROACETYLATIVE DIAZO GROUP TRANSFER:

(E)-1-DIAZO-4-PHENYL-3-BUTEN-2-ONE

(3-Buten-2-one, 1-diazo-4-phenyl-)

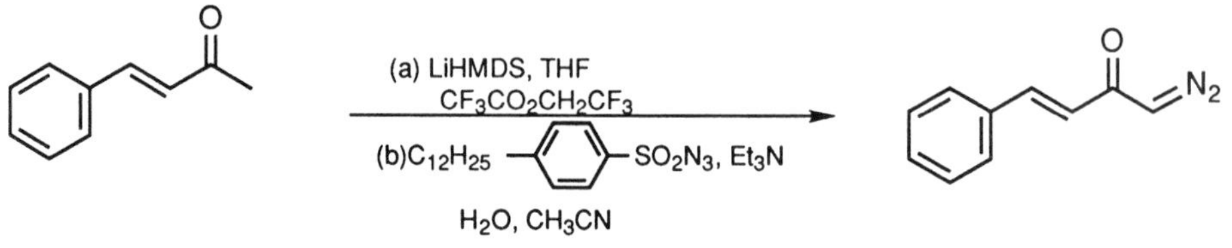

Submitted by Rick L. Danheiser, Raymond F. Miller, and Ronald G. Brisbois.[1]

Checked by Cameron Clark and Stephen F. Martin.

1. Procedure

Caution! Diazo compounds are presumed to be toxic and potentially explosive and therefore should be handled with caution in a fume hood. Although in carrying out this reaction numerous times we have never observed an explosion, we recommend that this preparation be conducted behind a safety shield.

A 500-mL, three-necked, round-bottomed flask is equipped with a mechanical stirrer, nitrogen inlet adapter, and 150-mL pressure-equalizing dropping funnel fitted with a rubber septum (Note 1). The flask is charged with 70 mL of dry tetrahydrofuran (Note 2) and 15.9 mL (0.075 mol) of 1,1,1,3,3,3-hexamethyldisilazane (Note 3), and then cooled in an ice-water bath while 28.8 mL (0.072 mol) of a 2.50 M solution of n-butyllithium in hexane (Note 4) is added dropwise over 5-10 min. After 10 min, the resulting solution is cooled at -78°C in a dry ice-acetone bath, and a solution of 10.0 g (0.068 mol) of trans-4-phenyl-3-buten-2-one (Note 5) in 70 mL of dry tetrahydrofuran is added dropwise over 25 min. The dropping funnel is washed with two 5-mL portions of tetrahydrofuran and then replaced with a rubber septum. The yellow reaction mixture is allowed to stir for 30 min at -78°C, and then 10.1 mL (0.075 mol) of 2,2,2-trifluoroethyl trifluoroacetate (TFEA, Note 6) is added rapidly in one portion via syringe (over ~ 5 sec). After 10 min, the reaction mixture is poured into a 1-L separatory funnel containing 100 mL of diethyl ether and 200 mL of 5% aqueous hydrochloric acid. The aqueous layer is separated and extracted with 50 mL of diethyl ether. The combined organic layers are washed with 200 mL of saturated sodium chloride solution, dried over anhydrous sodium sulfate, filtered, and concentrated under reduced pressure using a rotary evaporator to afford 18.61 g of a yellow oil. This yellow oil is immediately dissolved in 70 mL of acetonitrile (Note 7) and transferred to a 500-mL, one-necked flask equipped with a magnetic stirring bar and a 150-mL pressure equalizing dropping funnel fitted with a nitrogen inlet adapter. Water (1.2 mL, 0.069 mol), triethylamine (14.3 mL, 0.103 mol) (Note 8), and a solution of 4-dodecylbenzenesulfonyl azide[2] (35.74 g, 0.103 mol) (Note 9) in 10 mL of acetonitrile are then sequentially added (each over ~ 1-2 min) via the dropping funnel. The resulting yellow solution is allowed to stir at room temperature for 6.5 hr and then is poured into a 1-L separatory funnel containing 100 mL of diethyl ether and 200 mL of aqueous 5% sodium hydroxide (NaOH). The organic layer is separated, washed successively with three 200-mL portions of 5% aq NaOH, four 200-mL portions of water, 200 mL of saturated sodium chloride, dried over anhydrous sodium sulfate, filtered, and concentrated at reduced pressure using a rotary evaporator to yield 23.17 g of crude reaction product as a light brown oil. The crude reaction product is purified by column chromatography on 230-400 mesh silica gel (30 times by weight, elution with 5-10% diethyl ether-hexane) to furnish 9.54-9.80 g (81-83%) of (E)-1-diazo-4-phenyl-3-buten-2-one (mp 68-69°C) as a bright yellow solid (Notes 10, 11).

2. Notes

1. The apparatus is flame-dried under reduced pressure and then maintained under an atmosphere of nitrogen during the course of the reaction.

2. Tetrahydrofuran was distilled from sodium benzophenone ketyl immediately before use.

3. 1,1,1,3,3,3-Hexamethyldisilazane was purchased from Aldrich Chemical Company, Inc., and was distilled from calcium hydride prior to use.

4. n-Butyllithium was purchased from Aldrich Chemical Company, Inc., and was titrated prior to use according to the the method of Watson and Eastham.[3]

5. trans-4-Phenyl-3-buten-2-one was purchased from Aldrich Chemical Company, Inc., and used without further purification.

6. 2,2,2-Trifluoroethyl trifluoroacetate was purchased from Aldrich Chemical Company, Inc., and used without further purification.

7. Acetonitrile was distilled from calcium hydride immediately prior to use.

8. Triethylamine was purchased from Fisher Chemical Company and distilled from calcium hydride before use.

9. The submitters originally used methanesulfonyl azide,[4] but the Board of Editors of *Organic Syntheses* requested substitution of the much less shock sensitive reagent 4-dodecylbenzenesulfonyl azide. The use of methanesulfonyl azide has previously been recommended,[4a] since excess reagent as well as certain formamide by-products can be easily separated from the desired diazo ketone product during workup by extraction into dilute aqueous base.

10. The product has the following spectral properties: IR (CCl_4) cm^{-1}: 3150-3000, 2090, 1645, 1600, 1445, 1360, 1180, 1140, 1095, 1070, 970, 690; ^1H NMR (300 MHz, $CDCl_3$) δ: 5.54 (s, 1 H), 6.60 (d, 1 H, J = 15.8), 7.30-7.34 (m, 3 H), 7.46-7.49 (m, 2 H), 7.57 (d, 1 H, J = 15.8) ; ^{13}C NMR (75 MHz, $CDCl_3$) δ: 55.8, 123.5, 127.8,

128.5, 129.9, 134.0, 140.1, 184.0; Anal. Calcd for $C_{10}H_8N_2O$: C, 69.76; H, 4.68; N, 16.27. Found: C, 69.65; H, 4.84; N, 16.32.

11. When 4-dodecylbenzenesulfonyl azide is used for the diazo transfer reaction, the crude reaction product is contaminated with by-products that cannot be separated during basic workup, and consequently column chromatography is required for the purification of the diazo ketone. Use of mesyl azide for the diazo transfer reaction allows purification of the crude reaction product by recrystallization from diethyl ether-pentane to obtain 10.11 g (86%) of the desired diazo ketone.

Waste Disposal Information

All toxic materials were disposed of in accordance with "Prudent Practices for Disposal of Chemicals from Laboratories"; National Academic Press; Washington, DC, 1983.

3. Discussion

The importance of α-diazo ketones as synthetic intermediates has led to the development of a number of general methods for their preparation.[5] Particularly popular approaches include the acylation of diazo alkanes and the base-catalyzed "diazo group transfer" reaction of sulfonyl azides with β-dicarbonyl compounds.[6,7] While *direct* diazo transfer to ketone enolates is usually not a feasible process,[8,9] diazo transfer to simple ketones can be achieved in two steps by employing an indirect "deformylative diazo transfer" strategy in which the ketone is first formylated under Claisen condensation conditions, and then treated with a sulfonyl azide reagent such as p-toluenesulfonyl azide.[6a,6c,9,10,11]

Unfortunately, several important classes of α-diazo ketones cannot be prepared in good yield via these standard methods. α'-Diazo derivatives of α,β-unsaturated ketones, for example, have previously proved to be particularly difficult to prepare.[11b,12] The acylation of diazomethane with α,β-unsaturated acid chlorides and anhydrides is generally not a successful reaction because of the facility of dipolar cycloaddition to conjugated double bonds, which leads in this case to the formation of mixtures of isomeric pyrazolines. Also problematic are diazo transfer reactions involving base-sensitive substrates such as certain α,β-enones and heteroaryl ketones. Finally, the relatively harsh conditions and lack of regioselectivity associated with the thermodynamically controlled Claisen formylation step in the "deformylative" diazo transfer procedure limit the utility of this method when applied to the synthesis of diazo derivatives of many enones and unsymmetrical saturated ketones.

The *detrifluoroacetylative* diazo transfer procedure described here[13] is more general than the classical deformylative strategy, and as indicated in the Table, gives superior results when applied to a variety of ketone substrates. The new method has proved particularly valuable in the preparation of diazo derivatives of α,β-enones. In the case of saturated ketones such as 4-tert-butylcyclohexanone, both methods give comparable results, although the new procedure is more convenient to carry out, and has the advantage of providing a regioselective means of effecting diazo transfer to unsymmetrical ketones.

A key feature of the new procedure is the activation of the ketone starting material as the corresponding α-trifluoroacetyl derivative. To our knowledge, the use of TFEA to activate ketones in this fashion has not previously been reported, although Doyle has employed a similar strategy to achieve diazo transfer to a base sensitive N-acyloxazolidone derivative.[14] In our experience, TFEA has proved superior to other trifluoroacetylating agents [e.g. CF_3CO_2Et, $(CF_3CO)_2O$] for this transformation; the reaction of ketone enolates with this ester takes place essentially instantaneously at

138

-78°C. By contrast, the formylation of ketone enolates with ethyl formate is usually carried out using sodium hydride or sodium ethoxide as base and generally requires 12 to 48 hr at room temperature for complete reaction.

Only one equivalent of base is required for the trifluoroacetylation step; apparently the chelated tetrahedral intermediate is stable at -78°C and the β-dicarbonyl product is not generated until workup. Crucial to the success of the trifluoroacetylation reaction in some cases is the selection of lithium hexamethyldisilazide (LiHMDS) for the generation of the ketone enolate; under otherwise identical conditions diazo transfer to several aryl ketones proceeds in dramatically reduced yield when lithium diisopropylamide is employed as base.

In summary, the method described here provides an efficient and convenient route to a variety of α-diazo ketones including unsaturated derivatives that were not previously available by diazo transfer. α-Diazo ketones serve as key intermediates in a number of important synthetic methods including the Arndt-Eistert homologation, the photo-Wolff ring contraction strategy, and the carbenoid-mediated cyclopropanation reaction. We anticipate that improved access to α-diazo ketones will serve to enhance the utility of these valuable synthetic strategies.

1. Department of Chemistry, Massachusetts Institute of Technology, Cambridge, MA 02139. We thank the National Institutes of Health (GM 28273) for generous financial support. R. F. M. was supported in part by NIH training grant CA 09112.

2. Hazen, G. G.; Bollinger, F. W.; Roberts, F. E.; Russ, W. K.; Seman, J. J.; Staskiewicz, S. *Org. Synth.* **1994**, *73*, 144.

3. Watson, S. C.; Eastham, J. F. *J. Organomet. Chem.* **1967**, *9*, 165.

4.   (a) Taber, D. F.; Ruckle, R. E., Jr.; Hennessy, M. *J. Org. Chem.* **1986**, *51*, 4077; (b) Lowe, G.; Ramsay, M. V. *J. Chem. Soc., Perkin Trans. 1* **1973**, 479; (c) Stork, G.; Szajewski, R. P. *J. Am. Chem. Soc.* **1974**, *96*, 5787.

5.   For reviews of methods for the synthesis of α-diazo ketones, see (a) Regitz, M.; Maas, G. "Diazo Compounds: Properties and Synthesis"; Academic Press: New York; 1986; (b) Regitz, M. In "The Chemistry of Diazonium and Diazo Groups"; Patai, S., Ed.; Wiley: New York, 1978; Part 2, Chapter 17, pp. 751-820.

6.   For reviews of diazo group transfer, see (a) Regitz, M. *Angew. Chem., Int. Ed. Engl.* **1967**, *6*, 733; (b) Regitz, M. *Synthesis* **1972**, 351; (c) Chapter 13 of ref. 5a.

7.   Recent methods for diazo group transfer include: (a) Koskinen, A. M. P.; Munoz, L. *J. Chem. Soc., Chem. Commun.* **1990**, 652; (b) McGuiness, M.; Shechter, H. *Tetrahedron Lett.* **1990**, *31*, 4987; (c) Popic, V. V.; Korneev, S. M.; Nikolaev, V. A.; Korobitsyna, I. K. *Synthesis* **1991**, 195.

8.   Diazo transfer from 2,4,6-triisopropylphenylsulfonyl azide to the enolate derivatives of hindered cyclic ketones can be achieved by using phase transfer conditions: Lombardo, L.; Mander, L. N. *Synthesis* **1980**, 368.

9.   Evans and co-workers have reported successful diazo transfer from p-nitrobenzenesulfonyl azide (PNBSA) to the enolate derivatives of an N-acyloxazolidinone and a benzyl ester: (a) Evans, D. A.; Britton, T. C.; Ellman, J. A.; Dorow, R. L. *J. Am. Chem. Soc.* **1990**, *112*, 4011. However, we have not been able to achieve efficient diazo transfer to *ketone* enolates employing these conditions. For example, exposure of the lithium enolate of acetophenone to 1.2 equiv of PNBSA in THF at -78°C for 15 min gave α-diazoacetophenone in only 21% yield.

10.   (a) Regitz, M.; Menz, F. *Chem. Ber.* **1968**, *101*, 2622; (b) Hendrickson, J. B.; Wolf, W. A. *J. Org. Chem.* **1968**, *33*, 3610; (c) Rosenberger, M.; Yates, P.; Wolf,

W. *Tetrahedron Lett.* **1964**, 2285; (d) Regitz, M.; Rüer, J.; Liedhegener, A. *Org. Synth. Coll. Vol. VI* **1988**, 389.

11. Other indirect diazo transfer routes to α-diazo ketones have been reported involving initial activation of the ketone by benzoylation and acylation with diethyl oxalate: (a) Metcalf, B. W.; Jund, K.; Burkhart, J. P. *Tetrahedron Lett.* **1980**, *21*, 15; (b) Harmon, R. E.; Sood, V. K.; Gupta, S. K. *Synthesis* **1974**, 577.

12. For discussion and examples, see (a) pp 498-99 of ref. 4a; (b) Fink, J.; Regitz, M. *Synthesis* **1985**, 569; (c) Itoh, M.; Sugihara, A. *Chem. Pharm. Bull.* **1969**, *17*, 2105; (d) Regitz, M.; Menz, F.; Liedhegener, A. *Justus Liebigs Ann. Chem.* **1970**, *739*, 174; (e) Rosenquist, N. R.; Chapman, O. L. *J. Org. Chem.* **1976**, *41*, 3326 and references cited therein.

13. Danheiser, R. L.; Miller, R. F.; Brisbois, R. G.; Park, S. Z. *J. Org. Chem.* **1990**, *55*, 1959.

14. Doyle, M. P.; Dorow, R. L.; Terpstra, J. W.; Rodenhouse; R. A. *J. Org. Chem.* **1985**, *50*, 1663.

TABLE I
SYNTHESIS OF α-DIAZO KETONES

| Entry | α–Diazo Ketone | Diazo Transfer Procedure[a] (Isolated Yield, %) | |
		via Formylation	via Trifluoroacetylation
1		73	95
2		57	92
3		56	81
4		71	63, 90[b]
5		17	83[c], 86
6		45	87
7		44	84
8		68	61

[a]Diazo transfer reactions were carried out using methanesulfonyl azide unless otherwise indicated. [b]The yield is corrected for recovered propiophenone. [c]4-Dodecylbenzenesulfonyl azide was employed for this diazo transfer reaction.

Appendix

Chemical Abstracts Nomenclature (Collective Index Number) (Registry Number)

(E)-1-Diazo-4-phenyl-3-buten-2-one: 3-Buten-2-one, 1-diazo-4-phenyl (8,9); (24265-71-2)

1,1,1,3,3,3-Hexamethyldisilazane; Disilazane, 1,1,1,3,3,3-hexamethyl- (8); Silanamine, 1,1,1-trimethyl-N-(trimethylsilyl)- (9); (999-97-3)

Butyllithium: Lithium, butyl- (8,9); (109-72-8)

trans-4-Phenyl-3-buten-2-one: 3-Buten-2-one, 4-phenyl-, (E)- (8,9); (1896-62-4)

2,2,2-Trifluoroethyl trifluoroacetate: Acetic acid, trifluoro-, 2,2,2-trifluoroethyl ester (8,9); (407-38-5)

Acetonitrile (8,9); (75-05-8)

4-Dodecylbenzenesulfonyl azide: Benzenesulfonyl azide, 4-dodecyl- (10); (79791-38-1)

Methanesulfonyl azide (8,9); (1516-70-7)

4-DODECYLBENZENESULFONYL AZIDES

(Benzenesulfonyl azides, 4-dodecyl-)

Submitted by G. G. Hazen, F. W. Bollinger, F. E. Roberts, W. K. Russ, J. J. Seman, and S. Staskiewicz.[1]

Checked by Mark Spaller and Stephen F. Martin.

1. Procedure

Caution! Although the mixture of dodecylbenzenesulfonyl azides is the safest of a group of diazo transfer reagents,[2] one should keep in mind the inherent instability, shock sensitivity, and explosive power of azides. All users should exercise appropriate caution.

4-Dodecylbenzenesulfonyl chlorides. A 250-mL, three-necked, round-bottomed flask, equipped with a mechanical overhead stirrer, a Claisen adapter bearing an immersion thermometer, a pressure-equalizing addition funnel, and reflux condenser, is charged with a solution of 60.00 g (0.184 mol) of dodecylbenzenesulfonic acids (Note 1) and 8.64 mL of dimethylformamide (DMF) in 60 mL of hexane. Stirring is initiated while the mixture is heated to 70°C using a heating mantle, and 22.1 mL (36.24 g, 0.304 mol) of thionyl chloride (Note 2) is added at a rate to maintain controlled reflux (Note 3). The required addition time is about 1 hr. The dark solution is heated an additional 2 hr at 70°C and cooled to 40°C (Note 4). While still warm (~40°C), the mixture is transferred to a 250-mL separatory funnel, and the dark lower layer is separated from the hexane solution (Note 5). The hexane layer is cooled to 25°C and washed with 60 mL of aqueous 5% sodium bicarbonate solution (Note 6). The bicarbonate wash is back extracted with 36 mL of hexane and the combined hexane layers are treated with 3 g of carbon (Notes 7, 8) and stirred for 2 hr at 25°C. The carbon is removed by filtration and the cake is washed with three portions (12-mL each) of hexane. The combined hexane layers plus the hexane washes are used to prepare the azide.

4-Dodecylbenzenesulfonyl azides. A 500-mL, three-necked, round-bottomed flask fitted with a mechanical overhead stirrer is charged with the hexane solution from step A. To this solution is added a solution of 11.6 g (0.178 mol based on the total solids from the sulfonyl chlorides above) of sodium azide (NaN_3) in 100 mL of water and 2.0 g of phase transfer catalyst (Aliquat 336) (Note 9). Stirring is initiated, and the reaction progress is monitored by thin layer chromatography (Note 10). Approximately 4 hr at 25°C is required to complete the reaction. The two-phase mixture is transferred to a 500-mL separatory funnel and the aqueous layer is removed. The hexane layer is washed with 100 mL of aqueous 5% sodium bicarbonate solution and dried over 28 g of anhydrous sodium sulfate. The drying agent is removed by suction filtration, and the cake is washed with 20 mL of hexane. The concentration and purity of the 4-dodecylbenzenesulfonyl azides are best determined by evaporation of a small sample to an oil of constant weight with visible spectrophotometric assay for the azide (Note 11). The hexane solution of dodecylbenzenesulfonyl azides, when standardized as above (Note 11), can be used as obtained for most applications. However, if desired, *careful concentration of the hexane solution under reduced pressure at room temperature* affords 58.2-61.4 g (90-95%) of the oily mixture of

145

dodecylbenzenesulfonyl azides; corrected for the assay of the azides the yield is usually 95% (Notes 12 and 13).

2. Notes

1. Dodecylbenzenesulfonic acids, a 97% mixture of branched chain isomers, was purchased from Spectrum Chemical Mfg. Corp.

2. Reagent grade thionyl chloride from Fisher Scientific Co. was used.

3. An excess of thionyl chloride - dimethylformamide catalyst is used to prevent formation of sulfonic anhydrides. A stoichiometric amount of thionyl chloride gives a much reduced yield.

4. The progress of the reaction is monitored by thin layer chromatography. A 0.1-mL sample is removed, evaporated to dryness and dissolved in 2 mL of hexane. The solution is spotted on an Analtech silica GF plate (8 cm x 2.5 cm) and developed in hexane/methylene chloride (4/1). Visualization by UV light shows $R_f = 0.4$ for the sulfonyl chlorides.

5. If allowed to cool to 25°C, the dark layer may solidify, hampering the separation. This very acidic layer is the excess thionyl chloride/DMF complex. It should be handled with proper protection in a ventilated area To facilitate visual identification of the layers, the checkers added about 25 mL of hexane.

6. The pH of the bicarbonate wash is a reflection of the efficiency with which the dark lower layer has been removed. In the course of a dozen runs, this pH ranged from 5.5 to 7.1. If the pH of the wash is below 5.5, a second wash with bicarbonate is necessary.

7. Nuchar SA carbon from Westvaco Co. was used.

8. It is essential that the carbon treatment be carried out within a few hours. Experiments where this treatment was delayed for 16 hr invariably produced an azide

146

mixture of lower purity (85%) and lower yield (80%). Although the sulfonyl chlorides hydrolyze only slightly (1-2%) in wet hexane in 24 hr, that amount of sulfonic acids in the presence of the phase transfer agent catalyzes the hydrolysis of sulfonyl chlorides. After treatment with carbon, the hexane solution of sulfonyl chlorides can be stored for several weeks in the refrigerator with little or no adverse effect on the next step.

9. Aliquat 336 (tri-n-alkylmethylammonium chloride) was obtained from Aldrich Chemical Company, Inc. The material is a mixture of C_8 and C_{10} chains with C_8 predominating. There is a slight initial exothermic reaction on adding the phase transfer catalyst. Intermittent cooling with a cold water bath is required to keep the temperature below 35°C.

10. A sample of the hexane layer from the reaction mixture is diluted 10 fold with hexane, spotted and developed as described in Note 4. Visualization by UV light shows $R_f = 0.3$ for sulfonyl azides.

11. The assay for sulfonyl azides is adapted from the method of Siewinski, et al.[3] The azide content of the hexane solution was assayed as follows to determine the contained yield:

1. Standard Curve:

Stock solution: 80 mg of NaN_3 diluted to the mark in a 100-mL volumetric flask with 0.1 N NaOH-MeOH.

Procedure: Into a series of four, 100-mL volumetric flasks, transfer 5, 7, 10 and 15 mL respectively of the stock solution. Into a 100-mL volumetric labelled blank, pipet 10 mL of 0.1 N NaOH-MeOH solution. To all add 2 drops of 0.1% ethanolic phenolphthalein indicator solution, and 20 mL of aqueous 1.5% sodium sulfate (Na_2SO_4). Acidify each, in turn, to the phenolphthalein end point with 1 N hydrochloric acid and immediately add 25 mL of 1 M ferric ammonium sulfate [$Fe(NH_4)(SO_4)_2$] solution. Dilute to the mark with 1.5% Na_2SO_4. Let stand 10 min, then read absorbance at 458 nm. Plot absorbance vs. concentration.

2. Sample:

Pipet 5 mL of the hexane solution containing 4-dodecylbenzenesulfonyl azides into a 100-mL volumetric flask and dilute to the mark with methanol. Pipet 5 mL of this solution into a small stoppered flask. Add 2 mL of aqueous 1 N potassium hydroxide solution and heat at 75°C for ~20 min. Allow to cool to room temperature, add 2 drops of 0.1% phenolphthalein solution, and 10 mL of 1.5% Na_2SO_4. Shake, then transfer quantitatively to a 60-mL separatory funnel. Add 10 mL of butanol (or isoamyl alcohol) to the sample flask, shake, then transfer to the separatory funnel. Shake the funnel, let the layers separate, then remove the bottom (H_2O) layer into a 100-mL volumetric flask. Add an additional 10 mL of 1.5% Na_2SO_4 to the alcohol layer in the separatory funnel, shake, let the layers separate, then transfer the water layer to the volumetric flask. Neutralize the combined water layers to the phenolphthalein end point with 1 N hydrochloric acid, then immediately add 25 mL of $Fe(NH_4)(SO_4)_2$. Dilute to the mark with 1.5% Na_2SO_4 solution, let stand 10 min, then read absorbance at 458 nm. Read azide concentration against the NaN_3 calibration curve.

12. The checkers determined the yield by evaporation of the hexane solution to constant weight (3-10 hr at 0.1 mm). The yields cited are based on the assumption that the 3% impurity in the starting sulfonic acids is not present in the final product. The checkers found the material obtained upon concentration, to be sufficiently pure for use without further purification.

13. The spectroscopic data for the mixture of four isomeric secondary dodecylbenzenesulfonyl azides (~2.5:1.6:1.6:1.0) is as follows: ^{1}H NMR (400 MHz, $CDCl_3$) δ: 0.70-1.00 (m, 6H), 1.00-1.50 (m, 12H), 1.50-1.80 (m, 6H), 2.90-2.50 (m, 1H), 7.25-7.45 (m 2H), 7.87 (m 2H); ^{13}C NMR (100 MHz, $CDCl_3$) δ: 12.07, 13.94, 14.02, 14.05, 14.07, 14.10, 20.63, 21.89, 22.50, 22.61, 22.63, 22.65, 22.69, 27.18, 27.48, 27.52, 27.58, 29.15, 29.26, 29.30, 29.32, 29.44, 29.46, 29.49, 29.51, 29.57, 29.61, 29.72, 31.69, 31.82, 31.83, 31.88, 31.91, 36.21, 36.34, 36.60, 36.61, 36.64, 38.06,

38.85, 40.24, 46.11, 46.38, 48.15, 127.55, 127.66, 128.33, 128.93, 129.01, 135.82, 135.85, 154.56, 154.82, 154.87, 156.09; IR (film) 2126 cm[-1].

Waste Disposal Information

All toxic materials were disposed of in accordance with "Prudent Practices for Disposal of Chemicals from Laboratories"; National Academic Press; Washington, DC, 1983.

3. Discussion

4-Dodecylbenzenesulfonyl chlorides have been prepared from the corresponding acids using chlorosulfonic acid,[4] phosphorus oxychloride,[2] and thionyl chloride.[5] The use of catalytic amounts of DMF in conjunction with thionyl chloride is based on the work of H. Bosshard, et al.[6] The insolubility of the DMF/thionyl chloride complex in the reaction solvent permits easy removal at the end of reaction. Extraction with dilute base removes the last trace of acids and the solution is pure enough for the next step.

The method described above for the preparation of the mixture of 4-dodecylbenzenesulfonyl azides is new and based on the work of Bollinger and Hazen.[5,7] Sulfonyl azides have been prepared by diazotizing substituted sulfonyl hydrazides,[8] and treating sulfonyl halides in methanol-water,[9] ethanol-water,[10] acetone,[2,5,7] or acetone-water solutions[11] with aqueous or solid sodium azide,[5,7,12] Use of phase transfer catalysis for the preparation of sulfonyl azides is new, simple and effective. It avoids solvent changes and permits isolation of a hexane solution of sulfonyl azides without concentration.

149

The use and advantages of 4-dodecylbenzenesulfonyl azides as a diazo transfer agent are fully discussed by Hazen, Weinstock, Connell, and Bollinger.[7] In contrast to p-toluenesulfonyl azide, that has the shock sensitivity of tetryl (N-methyl-N-2,4,6-tetranitroaniline) and the explosiveness of TNT, the mixture of 4-dodecylbenzenesulfonyl azides exhibits no shock sensitivity at the highest test level (150 kg cm) and 24% of the heat of decomposition measured in cal/g. p-Toluenesulfonyl azide appears as a diazo transfer agent in *Org. Synth., Coll. Vol. V* **1973**, 179; *VI*, **1988**, 389, 414 and its preparation is reported in the first of these. Two explosions during its preparation have been reported.[13,14]

1. Process Research Department, Merck Research Laboratories, Division of Merck & Co., Inc., Rahway, NJ 07065.

2. Bistline, R. G., Jr.; Noble, W. R.; Linfield, W. M. *J. Amer. Oil Chem. Soc.* **1974**, *51*, 126-132.

3. Siewinski, M.; Kubicz, Z.; Szewczuk, A. *Anal. Chem.* **1982,** *54*, 846-847.

4. Virnig, M. J. U.S. Patent 4 100 163, 1978; *Chem. Abstr.* **1978**, *89*, 215238g.

5. Bollinger, F. W.; Hazen, G. G. U.S. Patent 4 284 575, 1981; *Chem. Abstr.* **1982**, *96*, 19822y.

6. Bosshard, H. H.; Mory, R.; Schmid, M.; Zollinger, Hch. *Helv. Chim. Acta* **1959**, *42*, 1653-1658.

7. Hazen, G. G.; Weinstock, L. M.; Connell, R.; Bollinger, F. W. *Synth. Commun.* **1981**, *11*, 947-956.

8. Curtius, T.; Klavehn, W. *J. Prakt. Chem.* **1926**, *112*, 65-87.

9. Reagan, M. T.; Nickon, A.; *J. Am. Chem. Soc.* **1968**, *90*, 4096-4105.

10. Stout, D. M.; Takaya, T.; Meyers, A. I. *J. Org. Chem.* **1975**, *40*, 563-569.

11. Breslow, D. S.; Sloan, M. F.; Newburg, N. R.; Renfrow, W. B. *J. Am. Chem. Soc.* **1969**, *91*, 2273-2279.

12.   Boyer, J. H.; Mack, C. H.; Goebel, N.; Morgan, L. R., Jr. *J. Org. Chem.* **1958**, *23*, 1051-1053.

13.   Spencer, H. *Chem. Brit.* **1981**, *17*, 106.

14.   Rewicki, D.; Tuchscherer, C. *Angew. Chem., Int. Ed. Engl.* **1972**, *11*, 44-45.

Appendix

Chemical Abstracts Nomenclature (Collective Index Number);
(Registry Number)

4-Dodecylbenzenesulfonyl azide: Benzenesulfonyl azide, 4-dodecyl- (10);
[79791-37-1]

Dodecylbenzenesulfonic acid, 97%: Benzenesulfonic acid, dodecyl- (8,9);
(27176-87-0)

Dodecylbenzenesulfonyl chloride: Benzenesulfonyl chloride, dodecyl- (8,9);
(26248-27-1) or 4-Dodecyl- (9); (52499-14-6)

N,N-Dimethylformamide CANCER SUSPECT AGENT: Formamide, N,N-dimethyl- (8,9);
(68-12-2)

Thionyl chloride (8,9); (7719-09-7)

Sodium azide (8,9); (26628-22-8)

Aliquat 336: Ammonium, methyltrioctyl-, chloride (8); 1-Octanaminium, N-methyl-N,N-dioctyl-, chloride (9); (5137-55-3)

Sodium sulfate: Sulfuric acid disodium salt (9); (7757-82-6)

Ferric ammonium sulfate: Sulfuric acid, ammonium iron (3+) salt (2:1:1),
dodecahydrate (9); (7783-83-7)

BIS(TRIFLUOROETHYL) (CARBOETHOXYMETHYL)PHOSPHONATE

(Acetic acid, [bis(2,2,2-trifluoroethoxy)phosphinyl]-, ethyl ester)

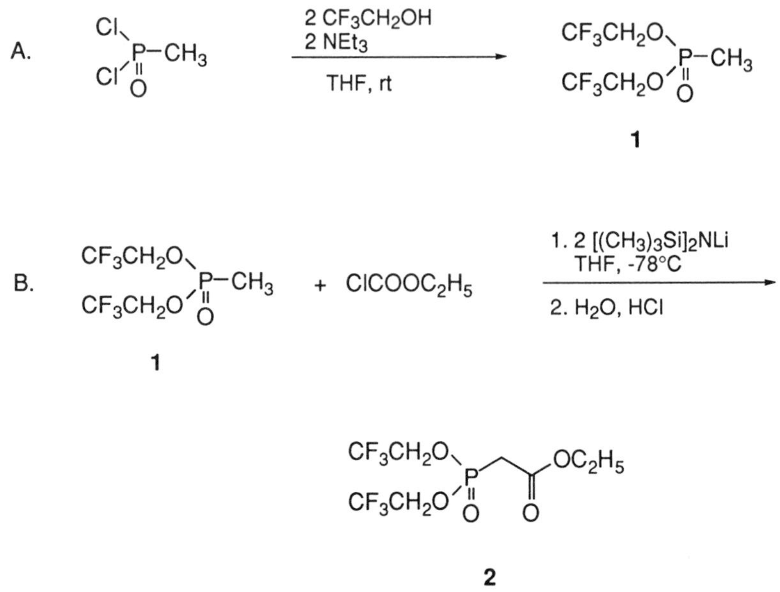

A.

$$CH_3\text{-}P(Cl)_2\text{=}O \quad \xrightarrow[\text{THF, rt}]{\begin{array}{c}2\ CF_3CH_2OH\\2\ NEt_3\end{array}} \quad (CF_3CH_2O)_2P(CH_3)\text{=}O$$

1

B.

$$(CF_3CH_2O)_2P(CH_3)\text{=}O \quad + \quad ClCOOC_2H_5 \quad \xrightarrow[\text{2. } H_2O,\ HCl]{\begin{array}{c}1.\ 2\ [(CH_3)_3Si]_2NLi\\\text{THF, -78°C}\end{array}}$$

1

$$(CF_3CH_2O)_2P(\text{=}O)CH_2COOC_2H_5$$

2

Submitted by Carl Patois[1], Philippe Savignac[1], Elie About-Jaudet,[2] and Noël Collignon.[2]

Checked by Andrzej R. Daniewski, Bryon K. Tilley, and David L. Coffen.

1. Procedure

A. Bis(trifluoroethyl) methylphosphonate **1**. All glassware is oven-dried. A 1-L, four-necked, round-bottomed flask is fitted with an efficient mechanical stirrer, a thermometer, reflux condenser with a bubbler, and a 200-mL, pressure-equalizing dropping funnel with a nitrogen inlet. Under a gentle flow of nitrogen, the flask is charged with 300 mL of anhydrous tetrahydrofuran (THF) (Note 1), 42 g (0.42 mol) of trifluoroethanol (Note 2), and 42.4 g of triethylamine (Note 2). Stirring is started, and the flask is immersed in a cool (~ 10°C) water bath. The dropping funnel is charged with a solution of 26.6 g (0.2 mol) of methylphosphonic dichloride (Note 3) in 100 mL of THF, that is subsequently added at a constant rate over 30 min. The internal temperature rises to 20-30°C, and a white solid precipitates. After the addition is complete, the water bath is removed, and the resulting mixture is stirred at room temperature for 2 hr. The organic salts are removed by suction filtration through a glass funnel, and the filter cake is washed with three 50-mL portions of THF. The solvent is completely removed under reduced pressure on a rotary evaporator. The residue is dissolved in 100 mL of dry ether and filtered to remove insoluble triethylamine hydrochloride, using an additional 40 mL of ether to aid the filtration/transfer. Solvent is again completely evaporated under reduced pressure. Crude product **1**, thus obtained, is transferred to a 100-mL pear-shaped flask, fitted with a short distillation column (Note 4), and distilled under reduced pressure to afford 45.3-48.0 g (89-93%) of pure bis(trifluoroethyl) methylphosphonate (**1**) as a colorless liquid, bp 88-91°C (14 mm), which affords white crystals on standing in the freezer (Note 5). When sufficiently pure, **1** will crystallize on standing at room temperature.

   B. Bis(trifluoroethyl) (carboethoxymethyl)phosphonate (**2**). An oven-dried, 1-L, four-necked, round-bottomed flask is fitted as above (Part A), flushed with nitrogen, and charged with 131 mL (0.210 mol) of a solution of butyllithium (1.6 M in hexane) (Note 6). The solution is cooled with stirring to approximately -20°C in a dry ice-acetone bath, and the dropping funnel is charged with a solution of 34 g (0.212 mol) of hexamethyldisilazane (Note 7) in 110 mL of tetrahydrofuran (THF) (Note 1). The contents of the addition funnel are added dropwise over 15 min during which time the yellow color practically disappears. A fine white suspension forms when the resulting solution is cooled to -78°C. A solution of 26 g (0.1 mol) of **1** and 11.5 g (0.106 mol) of ethyl chloroformate (Note 7) in 60 mL of THF is then added dropwise over 15 min via the addition funnel. During the course of the addition, the temperature rises to between -75° and -65°C, and the solution becomes clear. The reaction mixture is stirred at -70°C for an additional 15 min; the flask containing the reaction mixture is securely stoppered and stored overnight at -20°C in a freezer (Note 8). The cold (-20°C) reaction mixture is poured into a stirred mixture of 170 mL of 2 M hydrochloric acid and an equal volume of crushed ice, and 150 mL of methylene chloride. The organic layer is separated, and the aqueous layer is extracted twice with 50 mL of methylene chloride (Note 9). The combined organic extracts are diluted with 200 mL of hexane to facilitate drying, and then dried for ~12 hr over anhydrous magnesium sulfate (Note 10). After removal of the magnesium sulfate by filtration, the solvents are evaporated under reduced pressure using a rotary evaporator to give crude **2** as a pale yellow liquid. Subsequent distillation of this crude material in a short path distillation apparatus (Note 11) under reduced pressure gives, after a small forerun (Note 12), 25.5-26.3 g (77-79%) of bis(trifluoroethyl) (carboethoxymethyl) phosphonate (**2**) as a colorless oil, bp 88-97°C (0.04 mm) (Note 13).

2. Notes

1. Tetrahydrofuran available from SDS Company was purified by distillation from sodium and benzophenone.

2. The submitters used 99+% 2,2,2-trifluoroethanol, and 99% triethylamine, purchased from Janssen Chimica, without purification. The checkers used the same grade materials purchased from Aldrich Chemical Company, Inc., and Eastman Kodak, respectively.

3. Methylphosphonic dichloride (98%) can be purchased from Aldrich Chemical Company, Inc., or prepared according to a reported procedure.[3]

4. For the distillation, the submitters used a 14-cm fractional distillation column equipped with an 8-cm condenser.

5. The product displays the following spectroscopic data: ^{31}P (CDCl$_3$) δ: +35.0; ^{1}H (CDCl$_3$) δ: 1.68 (d, 3 H, J = 18.2), 4.35-4.42 (m, 4 H); ^{13}C (CDCl$_3$) δ: 10.9, 61.8, 122.6; IR (film) cm^{-1} : 1415 1317, 1292, 1258, 1186, 1169, 1111, 1075, 966, 922, 901, 845.

6. The submitters used 140 mL of 1.5 M butyllithium in hexane available from Janssen Chimica, standardized before use by titration against a solution of benzyl alcohol in toluene and 2,2'-biquinoline. The checkers used material from Aldrich Chemical Company, Inc.

7. The submitters used 98% 1,1,1,3,3,3-hexamethyldisilazane and 99% ethyl chloroformate purchased from Janssen Chimica and used without purification.

8. *Yields are low and erratic if the overnight cold storage is omitted!*

9. Prolonged exposure to work-up conditions should be avoided to prevent hydrolysis at phosphorus at this step.

10. In some cases, the organic extracts are contaminated with a salt resulting from partial hydrolysis at phosphorus. The salt precipitates slowly overnight and is removed together with the magnesium sulfate by filtration prior to the removal of the solvents.

11. A Büchi GKR-50 Kugelrohr distillation apparatus with three flasks was used by the submitters. The flask containing the crude product was in the upper part of the oven, and the collecting flask was just outside.

12. The forerun consisted mainly of siloxanes and other by-products.

13. The product displays the following NMR and physical data: ^{31}P (CDCl$_3$) δ: +24.0; ^{1}H (CDCl$_3$) δ: 1.29 (t, 3 H, J = 7.1), 3.14 (d, 2 H, J = 21.2), 4.21 (q, 2 H, J = 7.2), 4.45 (m, 4 H); ^{13}C (CDCl$_3$) δ: 14.1, 34.2, 62.8, 62.5, 122.7, 164.7. Anal. Calcd for

$C_8H_{11}F_6O_5P$: C, 28.93; H, 3.34; F, 34.32; P, 9.33. Found: C, 29.10; H, 3.49; F, 34.20; P, 9.16.

Waste Disposal Information

All toxic materials were disposed of in accordance with "Prudent Practices for Disposal of Chemicals from Laboratories"; National Academic Press; Washington, DC, 1983.

3. Discussion

Ethyl bis(trifluoroethyl) phosphonoacetate was introduced to synthesis by Still and Gennari in 1983.[4] It is a very useful reagent for the preparation of Z-acrylates by means of a Wittig-Hörner olefination. The procedure described here illustrates a general method for the one-pot formation of phosphonocarboxylates and enolates. The key step in the sequence is the formation of an α-lithiomethyl phosphonate in the presence of two equivalents of a hindered lithium amide, which promotes the direct generation of the enolate.[5] The present procedure has several advantages: 1) easy availability of reagents, 2) the possibility to extend the reaction to various carboxylate groups by the choice of the chloroformates (e.g., $ClCO_2R^3$; R^3 = Me, Et, i-Pr, tert-Bu), 3) the possibility to extend the reaction to α-alkylated phosphonocarboxylates by the choice of the starting alkylphosphonate $(R^1O)_2P(O)CH_2R^2$ (R^2 = H, Me, Et, Pr, etc.), and 4) high yields of the phosphonates, simplicity, and ease of isolation of the products.

$$\underset{R^1O}{\overset{R^1O}{\diagdown}}\!\!\!\!\underset{\overset{\|}{O}}{\overset{R^2}{\underset{}{P{-}CH_2}}} \quad + \quad ClCOOR^3 \quad \xrightarrow[\text{2. }H_3O^+]{\text{1. 2 eq. lithium amide}} \quad \underset{R^1O}{\overset{R^1O}{\diagdown}}\!\!\!\!\underset{\overset{\|}{O}}{\overset{R^2}{\underset{}{P}}}\!\!\!\!\underset{\overset{\|}{O}}{OR^3}$$

1. Hétéroatomes et Coordination, URA CNRS 1499, DCPH, Ecole Polytechnique, 91128 Palaiseau CEDEX, France.

2. Laboratoire des Composés Organophosphorés INSA-IRCOF, BP 08, 76131 Mont-Saint-Aignan CEDEX, France.

3. Maier, L. *Phosphorus, Sulfur Silicon Relat. Elem.* **1990**, *47*, 465.

4. Still, W. C.; Gennari, C. *Tetrahedron Lett.* **1983**, *24*, 4405

5. Aboujaoude, E. E.; Collignon, N.; Teulade, M.-P.; Savignac, P. *Phosphorus Sulfur* **1985**, *25*, 57; Aboujaoude, E. E.; Lietjé, S.; Collignon, N.; Teulade, M. P.; Savignac, P. *Tetrahedron Lett.* **1985**, *26*, 4435 ; Tay, M. K.; About-Jaudet, E.; Collignon, N.; Teulade, M. P.; Savignac, P. *Synth. Commun.* **1988**, *18*, 1349.

Appendix
Chemical Abstracts Nomenclature (Collective Index Number); (Registry Number)

Bis(trifluoroethyl) (carboethoxymethyl)phosphonate: Acetic acid, [bis(2,2,2-trifluoroethoxy)phosphinyl]-, ethyl ester (12); (124755-24-4)

Bis(trifluoroethyl) methylphosphonate: Phosphonic acid, methyl-, bis(2,2,2-trifluoroethyl) ester (9); (757-95-9)

Trifluoroethanol: Ethanol, 2,2,2-trifluoro- (8,9); (75-89-8)

Triethylamine (8); Ethanamine, N,N-diethyl- (9); (121-44-8)

Methylphosphonic dichloride: Phosphonic dichloride, methyl- (9); (676-97-1)

Butyllithium: Lithium, butyl- (8,9); (109-72-8)

1,1,1,3,3,3-Hexamethyldisilazane: Disilazane, 1,1,1,3,3,3-hexamethyl- (8);

157

Silanamine, 1,1,1-trimethyl-N-(trimethylsilyl)- (9); (999-97-3)

Ethyl chloroformate: Formic acid, chloro-, ethyl ester (8); Carbonochloridic acid, ethyl ester (9); (541-41-3)

(+)-(2R,8aR*)-[(8,8-DIMETHOXYCAMPHORYL)SULFONYL]OXAZIRIDINE AND (+)-(2R,8aR*)-[(8,8-DICHLOROCAMPHORYL)-SULFONYL]OXAZIRIDINE

([4H-4a,7-Methanooxazirino[3,2-i][2,1]benzisothiazole, tetrahydro-8,8-dimethoxy-9,9-dimethyl-, 3,3-dioxide], [2R-(2α,4aα,7α,8aR*)]- and [4H-4a,7-Methanooxazirino[3,2-i][2,1]benzisothiazole, 8,8-dichlorotetrahydro-9,9-dimethyl-, 3,3-dioxide, [2R-(2α,4aα,7α,8aR*)]])

Submitted by Bang-Chi Chen,[1] Christopher K. Murphy,[1] Anil Kumar,[1] R. Thimma Reddy,[1] Charles Clark,[1] Ping Zhou,[1] Bryan M. Lewis,[1] Dinesh Gala,[2] Ingrid Mergelsberg,[3] Dominik Scherer,[3] Joseph Buckley,[2] Donald DiBenedetto,[2] and Franklin A. Davis.[1,4]
Checked by Thanh H. Nguyen and Albert I. Meyers.

1. Procedure

A. (-)-(3-Oxocamphorylsulfonyl)imine. A 2-L, single-necked, round-bottomed flask, equipped with a condenser and magnetic stirring bar, is charged with 42.6 g (0.2 mol) of (-)-(camphorylsulfonyl)imine (Note 1), 30.0 g (0.27 mol) of selenium dioxide (Note 2), and 500 mL of reagent grade acetic acid (Note 3). The mixture is stirred at reflux for 14 hr (Note 4), and the black selenium metal that separates is removed by suction filtration of the hot reaction mixture using a 250-mL porcelain filter funnel. The funnel, flask and residue are washed with 50 mL of acetic acid. Removal of solvent using a rotary evaporator gives 43.0-47.0 g of crude (-)-(3-oxocamphorylsulfonyl)imine as a dark orange solid that is dried under reduced pressure overnight in a desiccator (Notes 5 and 6). This product is suitable for further reaction without purification.

B. (+)-[(7,7-Dimethoxycamphoryl)sulfonyl]imine. A 500-mL, single-necked, round-bottomed flask, equipped with a condenser, magnetic stirring bar, and nitrogen inlet, is charged with 22.8 g (0.1 mol) of the crude (-)-(3-oxocamphorylsulfonyl)imine prepared above, 125 mL of trimethyl orthoformate, 68 mL of methanol, and 5 mL of concd sulfuric acid. The mixture is heated to 85-90°C in an oil bath (Notes 7 and 8). After heating for 4 hr, during which time precipitation of the product is observed, the reaction mixture is cooled to room temperature, and an additional 30 mL of trimethyl orthoformate and 10 mL of methanol are added. Heating is continued for an additional hour, after which the reaction mixture is cooled to room temperature and

160

transferred with the aid of 250 mL of methylene chloride to a 500-mL separatory funnel. The solution is washed successively with water (100 mL), aqueous 20% sodium bicarbonate solution (100 mL), water (4 x 100 mL), and brine (100 mL), and dried over anhydrous magnesium sulfate. Filtration and removal of the solvent under reduced pressure affords 22.8-23.0 g (83-85%) of the crude (+)-(7,7-dimethoxy camphorylsulfonyl)imine as a light pink solid, mp 179-184ºC (Notes 9 and 10). This product is suitable for further reaction without purification.

C.   *(+)-[(7,7-Dichlorocamphoryl)sulfonyl]imine.* A 1-L, three-necked, round-bottomed flask, equipped with a mechanical stirrer or magnetic stirring bar, thermometer, and 250-mL pressure-equalizing addition funnel with a nitrogen inlet, is placed under a nitrogen atmosphere and charged with a solution of 50 g (0.235 mol) of (-)-(camphorsulfonyl)imine (Note 1) in 250 mL of ethyl acetate. To this solution is added 71 g (0.47 mol) of 1,8-diazabicyclo[5.4.0]undec-7-ene (DBU) (Note 11) over 30 min, and the reaction mixture is stirred for 30 min at room temperature. To the resulting mixture is then added 51.5 g (0.26 mol) of 1,3-dichloro-5,5-dimethylhydantoin (Note 12) in portions over 90 min while maintaining the reaction temperature at 20-25°C by cooling with an ice-water bath. When the reaction is complete, typically 30-60 min as determined by HPLC or TLC (Notes 13 and 14), 400 mL of water is slowly added while keeping the temperature at 20-25°C. The pH of the reaction mixture is adjusted to 7-7.5 by the addition of ~ 25 mL of concd hydrochloric acid, and the ethyl acetate solvent is removed on the rotary evaporator at a maximum bath temperature of 60°C (Note 15). The resulting suspension is stirred for 1 hr at room temperature, then the solids are collected by suction filtration, washed with 500 mL of water, and dried in a draft oven at 50°C to a constant weight affording 62.0-63.0 g (94-95%) of white solid (+)-(7,7-dichlorocamphorylsulfonyl)imine (mp 170-175°C) (Notes 16 and 17). This product is suitable for further reaction without purification.

D. (+)-(2R,8aR*)-[(8,8-Dimethoxycamphoryl)sulfonyl]oxaziridine (**1**). In a 500-mL, three-necked, Morton flask, equipped with a mechanical stirrer with a Teflon stirring blade and stirrer bearing, a thermometer, and a 250-mL pressure-equalizing addition funnel, is charged with 20.9 g (0.077 mol) of crude (+)-(7,7-dimethoxy camphorylsulfonyl)imine, 175 mL of methylene chloride, and 1.6 g of Aliquat 336 (Note 18). The reaction mixture is cooled in an ice bath to 0°C, efficient stirring is initiated, and a solution of 63 g (0.38 mol) of potassium carbonate in 120 mL of water is added at 0-10°C followed by 27.2 g (0.114 mol) of 32% peracetic acid (Note 19) dropwise such that the temperature is maintained at 3-5°C. After the addition is completed, the reaction mixture is allowed to warm to room temperature and stirred until reaction is complete (typically 40 hr) as determined by ^{1}H NMR (Notes 4 and 20). On completion, 0.6 g of sodium sulfite is added, the reaction mixture is stirred for 30 min, and 5 mL of aqueous 30% sodium hydroxide is introduced. The reaction mixture is transferred to a 500-mL separatory funnel with the aid of 50 mL of methylene chloride, the phases are separated, and the aqueous phase is extracted twice with 50 mL of methylene chloride. The combined organic extracts are washed successively with a saturated solution of sodium bicarbonate (50 mL) and water (2 x 50 mL), and then dried over anhydrous magnesium sulfate. The solvents are removed under reduced pressure while maintaining the temperature below 40°C which affords an off-white solid that is suspended in 100 mL hexane and collected by suction. The solids are washed on the filter with two additional 100-mL portions of hexane to give 15.1 g of (+)-(2R,8aR*)-[(8,8-dimethoxycamphoryl)sulfonyl]oxaziridine (**1**) having mp 184-186°C (dec.) and $[\alpha]_D^{25}$ +91.6° (CHCl$_3$, *c* 3.39) (Note 21). A second crop is obtained by evaporating the filtrate to dryness and repeating the process which affords a total of 19.2-19.8 g (86-90%) of the oxaziridine (mp 184-186°C (dec.)). The product may be used as obtained for oxidations, but may be further purified, if desired, by recrystallization from 1400 mL of 95% ethanol.

162

This oxaziridine can also be prepared in ~ 4 hr by oxidation of the corresponding imine using 3-chloroperbenzoic acid (Note 22).

The antipode, (-)-(2S,8aR*)-[(8,8-dimethoxycamphoryl)sulfonyl]oxaziridine, (mp 189°C (dec.); $[\alpha]_D^{20}$ -91.3° (CHCl$_3$, c 0.5)) was prepared in a similar manner starting from (+)-(camphorylsulfonyl)imine.

E. *(+)-(2R,8aR*)-[(8,8-Dichlorocamphoryl)sulfonyl]oxaziridine* (**2**). A 1-L, three-necked, Morton flask, equipped with a mechanical stirrer with a Teflon stirring blade and stirrer bearing or a magnetic stirring bar, a thermometer, and a 250-mL pressure-equalizing addition funnel is charged with a solution of 48 g (0.17 mol) of (+)-[(7,7-dichlorocamphoryl)sulfonyl]imine in 300 mL of methylene chloride and 3.5 g of Aliquat 336 (Note 18). After cooling the resulting mixture to 0°C with an ice-bath, a solution of 119 g (0.86 mol) of potassium carbonate in 250 mL of water is added to the rapidly stirring reaction mixture while maintaining the temperature between 0-10°C. Then 45 g (0.188 mol) of 32% peracetic acid (Note 19) is added dropwise at such a rate (~ 30 min) that the temperature is maintained between 0-5°C. When the oxidation is complete, as determined by TLC (normally after stirring overnight) (Note 23), 1.3 g (0.099 mmol) of sodium sulfite is added between 0-10°C. After stirring for 30 min, 8 mL of 30% aqueous sodium hydroxide solution is added and the reaction mixture is warmed to room temperature. The organic phase is separated, the aqueous phase is extracted with methylene chloride (2 x 25 mL), and the combined organic phases are washed with saturated sodium bicarbonate solution (25 mL) and water (2 x 25 mL) (Note 24), and dried over anhydrous sodium sulfate. Approximately 210 mL of methylene chloride is evaporated from the organic phase under reduced pressure while maintaining the bath temperature below 40°C. During removal of the solvent crystallization occurs. The resulting slurry is diluted with 125 mL of hexane, cooled in an ice bath (0-5°C), and stirred for 1 hr. The product is collected by suction filtration, washed with hexane (100 mL), and dried at a maximum of 40°C in an air draft oven to yield 45.5-48.0 g (90-95%) of (+)-(2R,8aR*)-[(8,8-dichlorocamphoryl)sulfonyl]oxaziridine (2) (mp 182-186°C; $[\alpha]_D^{20}$ +91.4° (CHCl$_3$,c 0.5)) (Notes 25 and 26).

The antipode (-)-(2S,8aR*)-[(8,8-dichlorocamphoryl)sulfonyl]oxaziridine (mp 182-186°C; $[\alpha]_D^{20}$ -92.3° (CHCl$_3$, c 0.5)) was prepared in a similar manner starting from (+)-(camphorylsulfonyl)imine.

2. Notes

1. (-)-(Camphorylsulfonyl)imine may be purchased from Aldrich Chemical Company, Inc., or prepared according to reference 5.

2. Selenium dioxide was purchased from Aldrich Chemical Company, Inc.

3. Reagent grade acetic acid was purchased from Aldrich Chemical Company, Inc.

4. The progress of the reaction was monitored using ^{1}H NMR by observing the disappearance of the two methyl absorptions of (-)-(camphorylsulfonyl)imine at δ 0.88, and 1.09 in a sample obtained from a 0.5-mL aliquot which is concentrated to dryness on a rotary evaporator.

5. The crude product is of sufficient purity for the next step even if trace amounts of acetic acid and selenium dioxide are present. The material can be purified by crystallization from chloroform to provide product with mp 189-190°C (Lit.[6] mp 190-191°C) and $[\alpha]_D^{20}$ -178.5° (acetone, c 2.2).

6. The spectral properties of (-)-[(3-oxocamphoryl)sulfonyl]imine are as follows: ^{1}H NMR (300 MHz, CDCl$_3$) δ: 0.95 (s, 3 H), 1.13 (s, 3 H), 1.76-1.87 (m, 1 H), 1.92-2.02 (m, 1 H), 2.16-2.36 (m, 2 H), 2.74 (d, 1 H, J = 4.8), 3.20 (d, 1 H, J = 13.5), 3.42 (d, 1 H, J = 13.5); ^{13}C NMR (75 MHz, CDCl$_3$) δ: 18.41, 20.19, 22.28, 28.00, 44.65, 50.07, 59.04, 62.74, 181.38, 197.71; IR (KBr) cm^{-1}: 2950, 1761, 1654, 1340, 1166, 741.

7. Within 30 min of heating to 85-90°C, vigorous evolution of gas ensues.

8. The reaction was monitored using ^{1}H NMR by observing the disappearance of the two methyl absorptions at δ 0.95 and 1.13 of (-)-(3-oxocamphorylsulfonyl)imine.

9. This material can be purified by crystallization from absolute ethanol to give product with mp 186-7°C and $[\alpha]_D^{20}$ +7.2° (CHCl$_3$, c 3.6).

10. The spectral properties of (+)-[(7,7-dimethoxycamphoryl)sulfonyl]imine are as follows: ^{1}H NMR (300 MHz, CDCl$_3$) δ: 0.94 (s, 3 H), 1.04 (s, 3 H), 1.73-2.02 (m, 4 H), 2.29 (d, 1 H, J = 1.4), 2.93 (d, 1 H, J = 13.3), 3.12 (d, 1 H, J = 13.3), 3.30 (s, 3 H), 3.39 (s, 3 H); ^{13}C NMR (75 MHz, CDCl$_3$) δ: 20.4, 20.5, 20.6, 29.2, 46.0, 48.8, 50.3, 50.5, 52.0, 64.2, 103.0, 188.8; IR (KBr) cm^{-1}: 1620, 1340, 1160.

11. DBU was purchased from Fluka Chemie AG, Air Products, or Aldrich Chemical Company, Inc.

12. 1,3-Dichloro-5,5-dimethylhydantoin (DCDMH) was purchased from Aldrich Chemical Company and used without additional purification.

13. HPLC conditions were as follows: C$_{18}$-Novapak (Waters), 5μ; UV detector at 210 nm; mobile phase: acetonitrile/water (55/45) at a flow rate of 1 mL/min. Alternately, this reaction can be monitored by TLC: R$_f$ = 0.42 using CH$_2$Cl$_2$ and 10% molybdophosphoric acid in ethanol as the developer.

14. During the reaction, the suspended solids dissolve giving a clear solution.

15. In the event that foaming takes place a few drops of 2-octanol are added.

16. This solid can be purified by crystallization from 2-propanol to give product with mp 177-179°C, and $[\alpha]_D^{20}$ +7.9° (CHCl$_3$, c 2.1), $[\alpha]_{365}^{20}$ +97.8° (CH$_3$CN, c 1).

17. The spectral properties of (+)-[(7,7-dichlorocamphoryl)sulfonyl]imine are as follows: ^{1}H NMR (300 MHz, CDCl$_3$) δ: 1.12 (s, 3 H), 1.17 (s, 3 H), 1.75-1.93 (m, 1 H), 1.96-2.20 (m, 2 H), 2.23-2.38 (m, 1 H), 2.75 (d, 1 H, J = 3.5), 3.24 (d, 1 H, J = 13.5), 3.42 (d, 1 H, J = 13.5); ^{13}C NMR (75 MHz, CDCl$_3$) δ: 21.8, 25.1, 27.4, 47.8, 50.8, 61.2, 64.1, 81.9, 189.2.

18. Aliquat 336 (tricaprylylmethylammonium chloride) was purchased from Aldrich Chemical Company, Inc.

19. Peracetic acid, 32% in acetic acid, was purchased from Aldrich Chemical Company, Inc.

20. The reaction was monitored using ^{1}H NMR by observing the disappearance of the two methyl absorptions at δ 0.94 and 1.04 of (-)-(7,7-dimethoxycamphoryl sulfonyl)imine.

21. The spectral properties of (+)-(2R,8aR*)-[(8,8-dimethoxycamphoryl)-sulfonyl]oxaziridine are as follows: ^{1}H NMR (300 MHz, CDCl$_3$) δ: 1.02 (s, 3H), 1.28 (s, 3H), 1.70-1.95 (m, 4 H), 2.24 (d, 1 H, J = 4.0), 3.03 (d, 1 H, J = 14), 3.23 (s, 3 H), 3.24 (d, 1 H, J = 14), 3.30 (s, 3 H); ^{13}C NMR (75 MHz, CDCl$_3$) δ: 20.5, 21.6, 28.1, 45.1, 47.4, 50.5, 50.8, 52.9, 54.6, 97.6, 102.8; IR (KBr) cm^{-1}: 1356, 1165.

22. Biphasic basic oxidation using technical grade (50-60%) 3-chloroperbenzoic acid affords this oxaziridine in ~ 4 hr: In a 2-L, three-necked, Morton-flask equipped with a mechanical stirrer was placed 22.6 g (0.083 mol) of crude (+)-[(7,7-dimethoxycamphoryl)sulfonyl]imine, 42.6 g (0.13 mol) of 3-chloroperoxybenzoic acid (50-60%) in 450 mL of methylene chloride, and 450 mL of saturated potassium carbonate solution. The reaction mixture was stirred vigorously until the oxidation was complete as indicated by TLC (Note 27) at which time 500 mL of water was added, the organic layer was separated and the aqueous layer was extracted with methylene chloride (2 x 500 mL). The combined organic extracts were washed with saturated sodium sulfite (300 mL) and water (300 mL), and dried over anhydrous magnesium sulfate.

23. The reaction was monitored by TLC using silica gel plates (Kieselgel-60F, 254 nm, Merck), developing with CH$_2$Cl$_2$; for visualization spray with 5% molybdophosphoric acid in ethanol and heat.

24. The aqueous phases are monitored with Merckoquant 10011 test strips for their peroxide content prior to their disposal. The residual peroxide is neutralized (<1 ppm, the detection limit of the paper) by the addition of saturated sodium sulfite solution.

25. The spectral properties of (+)-(2R,8aR*)-[(8,8-dichlorocamphoryl)sulfonyl]-oxaziridine are as follows: ^{1}H NMR (300 MHz, CDCl$_3$) δ: 1.16 (s, 3 H), 1.48 (s, 3 H), 1.86-2.18 (m, 3 H), 2.30-2.40 (m, 1 H), 2.73 (d, 1 H, J = 3.9), 3.23 (d, 1 H, J = 14), 3.45 (d, 1 H, J = 14); ^{13}C NMR (75 MHz, CDCl$_3$) δ: 21.9, 23.31, 25.3, 26.8, 47.3, 49.4, 54.6, 62.5, 86.1, 99.1.

26. The product can be recrystallized from ethanol/ethyl acetate (1.5:1).

27. A 1-mL aliquot was removed from the organic layer, diluted with 2 mL of methylene chloride, and analyzed by TLC eluting with methylene chloride (I$_2$ visualization); imine R$_f$ = 0.34, oxaziridine R$_f$ = 0.51.

Waste Disposal Information

All toxic materials were disposed of in accordance with "Prudent Practices for Disposal of Chemicals from Laboratories"; National Academy Press: Washington, DC 1983.

Safety Information

In the Radex study, samples of oxaziridine 2 were heated in open glass tubes at 60°C/hr from ambient temperature to 260°C under atmospheric conditions. It was found that there is a strong exotherm with an onset at 165°-190°C (neat), at 135-158°C upon addition of stainless steel, and at 73°-84°C upon addition of FeCl$_3$·H$_2$O. Furthermore, the onset temperature in the FeCl$_3$·H$_2$O study was found to be dependent on the concentration of ferric ion. In each case as the exotherm occurred, each sample frothed violently out of the sample tube as a gas was produced.

3. Discussion

(-)-[(3-Oxocamphoryl)sulfonyl]imine has been prepared independently by Glahsl and Herrmann in 70% yield in a similar manner using dioxane as solvent, but the reaction required two weeks for completion.[7] The submitters observed that using acetic acid as the solvent dramatically reduces the time to ~ 14 hr and improves the yield to 90%.[6] (+)-[(7,7-Dimethoxycamphoryl)sulfonyl]imine has been prepared in a similar manner in 70% yield.[8] This procedure affords this material in 83-88% yield.[6] Chlorination of the aza enolate of (+)-(camphorylsulfonyl)imine, prepared by treatment with sodium bis(trimethylsilyl)amide, with N-chlorosuccinimide affords (+)-[(7,7-dichlorocamphoryl)sulfonyl]imine in 74% yield.[9] The procedure described here uses inexpensive DBU and 1,3-dichloro-5,5-dimethylhydantoin to give this imine in 95% yield and is applicable to large scale preparations.[10]

Oxidation of (+)-[(7,7-dimethoxycamphoryl)sulfonyl]imine (96%)[6] and (+)-[(7,7-dichlorocamphoryl)sulfonyl]imine (98%)[10] with 3-chloroperbenzoic acid has been reported. The procedure described here uses less hazardous and less expensive peracetic acid with the aid of Aliquat 336.[10] However, this system requires 40 hr vs. 4 hr using 3-chloroperbenzoic acid for oxidation of (+)-[(7,7-dimethoxy-camphoryl)sulfonyl]imine to the oxaziridine.

N-Sulfonyloxaziridines are an important class of selective, neutral, and aprotic oxidizing reagents.[11] Enantiopure N-sulfonyloxaziridines have been used in the asymmetric hydroxylation of enolates to enantiomerically enriched α-hydroxy carbonyl compounds,[9,11-13] the asymmetric oxidation of sulfides to sulfoxides,[14,15] selenides to selenoxides,[16] sulfenimines to sulfinimines,[17] and the epoxidation of alkenes.[18]

168

(+)-(2R,8aR*)-[(8,8-Dimethoxycamphoryl)sulfonyl]oxaziridine (**1**) and (+)-(2R,8aR*)-[(8,8-dichlorocamphoryl)sulfonyl]oxaziridine (**2**) are most effective for the hydroxylation of 2-substituted 1-tetralone enolates to 2-substituted 2-hydroxy-1-tetralones. The former oxaziridine, **1**, gives higher ee's (>90%) with tetralones having an 8-methoxy group, while the dichloro reagent **2** is more effective with those enolates lacking this substituent. For example (+)-**1** has been employed in highly enantioselective syntheses of (+)- and (-)-5,7-O-dimethyleucomol (>96% ee),[19] the AB-ring segments for γ-rhodomycinone and α-citromycinone (94% ee),[6] and (R)-(-)-2-acetyl-5,8-dimethoxy-1,2,3,4-tetrahydro-2-naphthol (>95% ee), a key intermediate in the synthesis of anthracyclinones.[20] The asymmetric synthesis of the AB ring of aklavinone (>95% ee)[21] and the homoisoflavanoids (+)-O-trimethylsappanone (94% ee) and (+)-O-trimethylbrazilin (92% ee) used oxaziridine (+)-**2**. Some representative examples using these reagents are given in the Table.

ASYMMETRIC HYDROXYLATION OF TETRALONE AND PROPIOPHENONE ENOLATES USING (CAMPHORYLSULFONYL)OXAZIRIDINES 1 AND 2

Oxaziridine	Ketone (X=H)	Base	Temp. (°C)	α–Hydroxy Tetralone (X=OH) %Yield[a]	%ee (Config.)	Ref.
(+)-1	2-methyl tetralone	NHMDS[a]	-78	66	36 (R)	6,9,21
(+)-2		NHMDS	-78	66	>95 (R)	6,9,21
(+)-1	5,8-dimethoxy-2-ethyl tetralone	NHMDS	-78	58	60 (R)	6,21
		LDA[b]	0	66	94 (R)	
(+)-2		LDA	-78	55	73 (R)	
(+)-1	5,7-dimethoxy-3-(4-methoxybenzyl)chroman-4-one	NHMDS	-78	75	77 (R)	19
		LDA	-78	72	≥96 (R)	
(+)-2		NHMDS	-78	66	88 (R)	
(+)-1	5,8-dimethoxy-2-(methoxycarbonyl) tetralone	NHMDS	0	73	56 (S)	20
		LDA	0	No Reaction		
		KHMDS[c]	-78	70	≥95 (S)	
(+)-2		NHMDS	0	63	47 (S)	

$$\text{Ph}\overset{\overset{\displaystyle O}{\|}}{C}\text{---}\overset{\displaystyle X}{CH}\text{---}CH_3$$

(+)-1	X	NHMDS	-78	73	79 (S)	6,9,21
(+)-2		NHMDS	-78	70	95 (S)	9

[a]Isolated yields. [b]NHMDS = Sodium bis(trimethylsilyl)amide. [b]LDA = Lithium diisopropylamide. [c]KHMDS = Potassium bis(trimethylsilyl)amide.

1. Department of Chemistry, Drexel University, Philadelphia, PA 19104.

2. Schering-Plough Research Institute, 2015 Galloping Hill Road, Kenilworth, NJ 07033.

3. Werthenstein Chemie AG, CH-6105 Schachen, Switzerland.

4. Author to whom correspondence should be addressed.

5. Towson, J. C.; Weismiller, M. C.; Lal, G. S.; Sheppard, A. C.; Davis, F. A. *Org. Synth.* **1990**, *69*, 158.

6. Davis, F. A.; Kumar, A.; Chen, B.-C. *J. Org. Chem.* **1991**, *56*, 1143.

7. Glahsl, G.; Herrmann, R. *J. Chem. Soc., Perkin Trans. I*, **1988**, 1753.

8. Verfürth, U.; Herrmann, R. *J. Chem. Soc., Perkin Trans. I*, **1990**, 2919.

9. Davis, F. A.; Weismiller, M. C.; Murphy, C. K.; Thimma Reddy, R.; Chen. B.-C. J. *Org. Chem.* **1992**, *57*, 7274.

10. Mergelsberg. I.; Gala, D.; Scherer, D.; DiBenedetto, D.; Tanner, M. *Tetrahedron Lett.*, **1992**, *33*, 161.

11. For reviews on the chemistry of N-sulfonyloxaziridines, see: a) Davis, F. A.; Sheppard, A. C. *Tetrahedron* **1989**, *45*, 5703; b) Davis, F. A.; Chen, B.-C. *Chem. Rev.*, **1992**, *92*, 919.

12. Davis, F. A.; Sheppard, A. C.; Chen, B.-C.; Haque, M. S. *J. Am. Chem. Soc.* **1990**, *112*, 6679.

13. Davis, F. A.; Weismiller, M. C. *J. Org. Chem.* **1990**, *55,* 3715.

14. Davis, F. A.; Thimma Reddy, R.; Weismiller, M. C. *J. Am. Chem. Soc.* **1989**, *111*, 5964.

15. Davis, F. A.; Thimma Reddy, R.; Han, W. Carroll, P. J. *J. Am. Chem. Soc.* **1992**, 114, 1428.

16. Davis, F. A.; Thimma Reddy, R. *J. Org. Chem.* **1992**, *57*, 2599.

17. Davis, F. A.; Thimma Reddy, R.; Reddy, R. E. *J. Org. Chem.* **1992**, *57*, 6387.

18. Davis, F. A.; Thimma Reddy, R.; McCauley, Jr., J. P; Przeslawski, R. M.; Harakal, M. E.; Carroll, P. J. *J. Org. Chem.* **1991**, *56,* 809.

19. Davis, F. A.; Chen, B.-C. *Tetrahedron Lett.* **1990**, *31*, 6823.

20. Davis, F. A.; Clark, C.; Kumar, A.; Chen, B.-C. *J. Org. Chem.* **1994**, *59*, 1184.

21. Davis, F. A.; Kumar, A.*Tetrahedron Lett.* **1991**, *3*

Appendix

Chemical Abstracts Nomenclature (Collective Index Number);

(Registry Number)

(+)-(2R,8aR*)-[(8,8-Dimethoxycamphoryl)sulfonyl]oxaziridine: 4H-4a,7-Methano-oxazirino[3,2-i][2,1]benzisothiazole, tetrahydro-8,8-dimethoxy-9,9-dimethyl-, 3,3-dioxide, [2R-(2α,4aα,7α,8aR*)]- (12); (131863-82-5)

(+)-(2R,8aR*)-[(8,8-Dichlorocamphoryl)sulfonyl]oxaziridine: 4H-4a,7-Methano-oxazirino[3,2-i][2,1]benzisothiazole, 8,8-dichlorotetrahydro-9,9-dimethyl-, 3,3-dioxide, [2R-(2α,4aα,7α,8aR*)]- (12); (127184-05-8)

(-)-[(3-Oxocamphoryl)sulfonyl]imine: 3H-3a,6-Methano-2,1-benzisothiazole-7(4H)-one, 5,6-dihydro-8,8-dimethyl-, 2,2-dioxide, (3aS)- (12); (119106-38-6)

(-)-(Camphorylsulfonyl)imine: 3H-3a,6-Methano-2,1-benzisothaizole, 4,5,6,7-tetrahydro-8,8-dimethyl-, 2,2-dioxide, (3aS)- (9); (60886-80-8)

Selenium dioxide: Selenium oxide (8,9); (7446-08-4)

Acetic acid (8,9); (64-19-7)

(+)-[(7,7-Dimethoxycamphoryl)sulfonyl]imine: 3H-3a,6-Methano-2,1-benzisothiazole, 4,5,6,7-tetrahydro-7,7-dimethoxy-8,8-dimethyl-, 2,2-dioxide, [3aS]- (12); (131863-80-4)

Trimethyl orthoformate: Orthoformic acid, trimethyl ester (8); Methane, trimethoxy- (9); (149-73-5)

(+)-[(7,7-Dichlorocamphoryl)sulfonyl]imine: 3H-3a,6-Methano-2,1-benzisothiazole, 7,7-dichloro-4,5,6,7-tetrahydro-8,8-dimethyl-, 2,2-dioxide, (3aS)- (12); (127184-04-7)

1,8-Dizazbicylco[5.4.0]undec-7-ene: DBU: Pyrimido[1,2-a]azepine, 2,3,4,6,7,8,9,10-octahydro- (8,9); (6674-22-2)

1,3-Dichloro-5,5-dimethylhydantoin: Hydantoin, 1,3-dichloro-5,5-dimethyl- (8); 2,4-Imidazolidinedione, 1,3-dichloro-5,5-dimethyl- (9); (118-52-5)

Aliquat 336: Ammonium, methyltrioctyl-, chloride (8); 1-Octanaminium, N-methyl-N,N-dioctyl-, chloride (9); (5137-55-3)

Peracetic acid: Peroxyacetic acid (8); Ethaneperoxoic acid (9); (79-21-0)

m-Chloroperbenzoic acid: Peroxybenzoic acid, m-chloro- (8); Benzenecarboperoxoic acid, 3-chloro- (9); (937-14-4)

(-)-(2S,8aR*)-[(8,8-Dimethoxycamphoryl)sulfonyl]oxaziridine: 4H-4a,7-Methanooxazirino[3,2-i][2,1]benzisothiazole, tetrahydro-8,8-dimethoxy-9,9-dimethyl-, 3,3-dioxide, [2S-(2α,4aα,7α, 8aR*)]- (12); (132342-04-2)

(+)-(Camphorylsulfonyl)imine: 3H-3a,6-Methano-2,1-benzisothiazole, 4,5,6,7-tetrahydro-8,8-dimethyl-, 2,2-dioxide, (3aR)- (12); (107869-45-4)

(-)-(2S,8aR*)-[(8,8-Dichlorocamphoryl)sulfonyl]oxaziridine: 4H-4a,7-Methanooxazirino[3,2-i][2,1]benzisothiazole, 8,8-dichlorotetrahydro-9,9-dimethyl-, 3,3-dioxide, [2S-(2α,4aα,7α,8aR*)]- (13); (139628-16-3)

ASYMMETRIC SYNTHESIS OF TRANS-2-AMINOCYCLOHEXANECARBOXYLIC ACID DERIVATIVES FROM PYRROLOBENZODIAZEPINE-5,11-DIONES

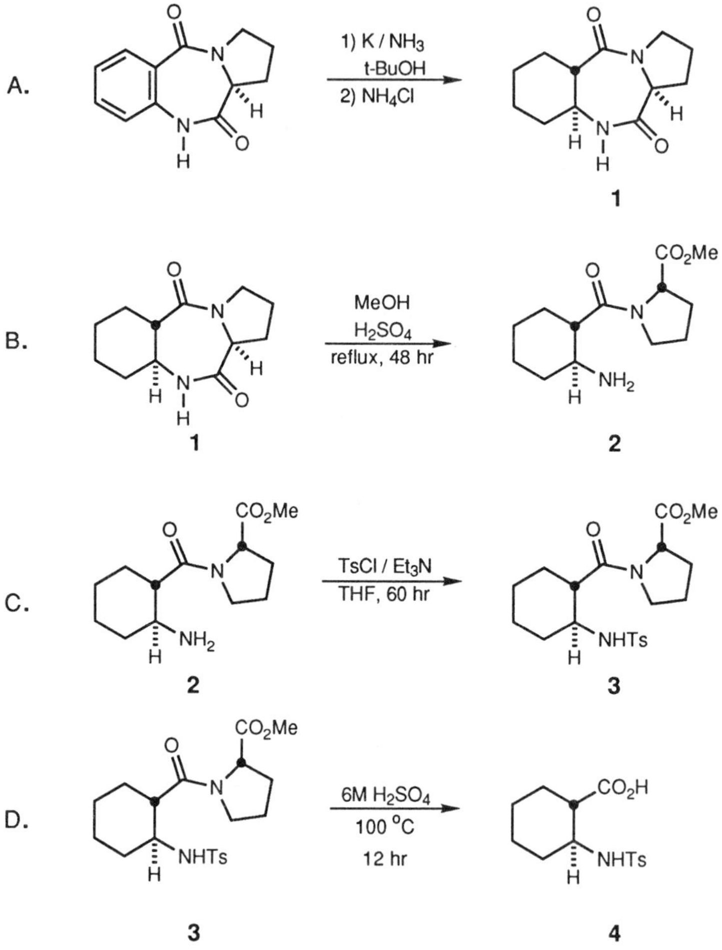

A.

1) K / NH$_3$
   t-BuOH
2) NH$_4$Cl

1

B.

1

MeOH
H$_2$SO$_4$
reflux, 48 hr

CO$_2$Me

NH$_2$

2

C.

CO$_2$Me

NH$_2$

2

TsCl / Et$_3$N
THF, 60 hr

CO$_2$Me

NHTs

3

D.

CO$_2$Me

NHTs

3

6M H$_2$SO$_4$
100 °C
12 hr

CO$_2$H

NHTs

4

Submitted by Arthur G. Schultz and Carlos W. Alva.[1]
Checked by Steven S. Henry and Albert I. Meyers.

1. Procedure

A. *(5aS,9aS,11aS)-Perhydro-5H-pyrrolo[2,1-c][1,4]benzodiazepine-5,11-dione*
(**1**). A 3-L, three-necked, round-bottomed flask, equipped with a mechanical stirrer bearing a glass paddle, a dry ice/acetone cooled cold-finger condenser bearing a nitrogen inlet/outlet valve vented through a mineral oil bubbler, and a gas inlet (Note 1) is placed under a nitrogen atmosphere, and charged with 25.0 g (0.116 mol) of 99% pure (S)-(+)-2,3-dihydro-1H-pyrrolo[2,1-c][1,4]benzodiazepine-5,11(10H,11aH)dione (Note 2) and a solution of 42.2 g (0.570 mol) of tert-butyl alcohol in 150 mL of dry tetrahydrofuran (THF) (Note 3). The mixture is cooled to -78°C (dry ice/acetone bath) and 2 L of dry ammonia is distilled into the mixture (Note 4). After stirring is initiated, a total of 35.7 g (0.912 mol) of potassium metal is added to the reaction mixture in small chunks at -78°C, and the resulting blue-colored solution is stirred for 1 hr. The reaction is then carefully quenched at -78°C by the addition of 61 g (1.14 mol) of solid ammonium chloride (Note 5). The cooling bath and condenser are removed, and the ammonia is allowed to evaporate overnight. The residue is suction filtered, and the filter cake is washed with 100 mL of chloroform. The filtrate is diluted with 200 mL of water, transferred to a 1-L separatory funnel, the organic phase is separated, and the aqueous phase is extracted three times with 100 mL of chloroform. The combined organic layers are dried over anhydrous sodium sulfate (Na_2SO_4), filtered, and concentrated under reduced pressure. The residue is crystallized from ethyl acetate, and the mother liquor purified by chromatography (Note 6) to give a total of 17.8 g (69%) of **1** (R_f 0.31, EtOAc/MeOH, 9:1) as a colorless solid, mp 225-227°C, $[\alpha]_D^{23}$ +55° (CHCl$_3$, c 0.61) (Notes 7 and 8).

B. *(1S,2S)-2-Amino-1-[((2S)-2-carbomethoxypyrrolidinyl)carbonyl]cyclohexane*
(**2**). A 250-mL, single-necked, round-bottomed flask equipped with a condenser bearing a nitrogen inlet /outlet valve as above (Part A) and a magnetic stirring bar, is

charged with a solution of 20.0 g (90.0 mmol) of **1**, 100 mL of anhydrous methanol, and 18.4 g (180 mmol) of concentrated sulfuric acid, placed under a nitrogen atmosphere, and heated at reflux (oil bath temperature at 80-85°C) for 48 hr or until TLC analysis indicates the disappearance of **1**. The solution is concd under reduced pressure, 100 mL of dichloromethane (CH_2Cl_2) is added, and the mixture is cooled to 0°C. A total of 150 mL of saturated aqueous sodium bicarbonate solution is added in small portions with good stirring, followed by 10-12 mL of concd ammonium hydroxide until a pH of 10 is reached (Note 9). After thorough mixing, the organic phase is separated and the aqueous phase is extracted twice with 50 mL of CH_2Cl_2. The combined organic phases are dried over Na_2SO_4, concentrated, and the resulting crude amine **2** used immediately in the next step (Note 10).

C. *(1S,2S)-2-(N-Tosylamino)-1-[((2S)-2-carbomethoxypyrrolidinyl)carbonyl] cyclohexane* (**3**). Approximately 22.9 g (~90 mmol) of crude amine **2** (prepared above in part B), 13.66 g (135 mmol) of triethylamine, and 100 mL of dry THF are placed in a 300-mL, round-bottomed flask, equipped with a pressure-equalizing dropping funnel, a magnetic stirring bar, and a nitrogen inlet. The dropping funnel is charged with a solution of 18.9 g (99.1 mmol) of p-toluenesulfonyl chloride (Note 11) in 50 mL dry THF. The reaction mixture is cooled to 0°C with magnetic stirring, and the solution of p-toluenesulfonyl chloride is delivered dropwise over a 30-min period. The resulting cloudy solution is stirred for 60 hr at ambient temperature. After this time period, the reaction mixture is diluted with 50 mL of saturated sodium chloride solution and 50 mL of ethyl acetate, transferred to a 500-mL separatory funnel, mixed thoroughly, and the organic phase separated. The aqueous phase is extracted twice with 50 mL of ethyl acetate. The combined organic layers are dried (Na_2SO_4), filtered, concentrated under reduced pressure, and the resulting residue purified by chromatography (Note 12) to give 22.43 g (61% from **1**) of **3** (R_f 0.34, $CHCl_3$/EtOAc, 1:1) as a colorless solid, mp 144-146 °C (Note 13).

D. *(1S,2S)-2-(N-Tosylamino)cyclohexanecarboxylic acid* (**4**). A 250-mL round-bottomed flask, equipped with a water-cooled condenser, is charged with 28.8 g (70.5 mmol) of **3** and 120 mL of 6 M sulfuric acid. The resulting heterogeneous mixture is heated at reflux (oil bath temperature 110-115°C) for 14 hr. After several hours at reflux, suspended solids are observed. The mixture is cooled to room temperature, transferred to a 500-mL separatory funnel, and extracted three times with 100 mL of CH_2Cl_2. The combined organic layers are dried over anhydrous magnesium sulfate, filtered, and concentrated under reduced pressure to afford **4** as a colorless solid. Recrystallization of the crude solid **4** from CH_2Cl_2 and ethyl acetate affords 12.16 g (58%) of pure **4** (R_f ~0.29, $CHCl_3$/EtOAc, 1:1) as colorless needles, mp 175-177°C, $[\alpha]_D^{20}$ +35° ($CHCl_3$, *c* 0.50) (Notes 14 and 15).

2. Notes

1. All glassware is flame dried and cooled under a stream of anhydrous nitrogen. All reactions are carried out under a positive pressure of nitrogen except for step D.

2. (S)-(+)-2,3-Dihydro-1H-pyrrolo[2,1-c][1,4]benzodiazepine-5,11(10H,11aH)–dione was purchased from the Aldrich Chemical Company, Inc. Other reagents, solvents, and drying agents were obtained from either Aldrich Chemical Company, Inc. or Fisher Scientific Company.

3. tert-Butyl alcohol is distilled prior to use. THF is distilled immediately before use from sodium benzophenone ketyl.

4. The ammonia is dried thoroughly over sodium metal before distillation into the reaction flask. If this step is not performed, a mixture of products may be obtained.[2]

5. Ammonium chloride is introduced as rapidly as possible, but with great care to avoid splashing and violent evaporation of the ammonia. The coloration of the reaction mixture will change from dark blue to colorless within 5-10 min.

6. As much product as possible is crystallized from the residue obtained from the organic extracts. The mother liquor from the crystallization is chromatographed on a column filled with neutral alumina (available from J. T. Baker Chemical Company, powder, Brockmann Activity Grade 1) using 20 g of alumina per gram of residue (elution with CHCl3/EtOAc, 1:1). Chromatography removes an impurity that makes crystallization difficult. Material purified by chromatography is also crystallized from ethyl acetate.

7. TLC analyses were performed on Macherey-Nagel Polygram SIL G UV/254 plates that were stained with a solution of phosphomolybdic acid in 95% ethanol.

8. Purified 1 has the following spectral data: ^{1}H NMR (300 MHz, CDCl3) δ: 1.18-1.43 (m, 4 H), 1.69-1.93 (m, 4 H), 1.95-2.08 (m, 2H), 2.19 (dt, 1 H, J = 11.3, 3.3), 2.43-2.53 (m, 1 H), 2.58 (m, 1 H), 3.46-3.69 (m, 3 H), 4.54 (t, 1 H, J = 6.8), 5.95 (s (br), 1 H); ^{13}H NMR (75 MHz, CDCl3) δ: 22.2, 25.0, 25.6, 27.9, 29.7, 32.4, 48.8, 51.7, 52.5, 56.2, 170.9, 171.2; IR (KBr) cm^{-1}: 3215, 1678, 1587; chemical ionization mass spectrum, m/z (relative intensity) M$^+$ + 1 (100). Anal. Calcd for C$_{12}$H$_{18}$N$_2$O$_2$: C, 64.85; H, 8.15. Found: C, 64.74; H, 8.14.

9. The aqueous layer must be strongly alkaline to enable extraction of the amine.

10. The free amine cyclizes to the diamide 1 upon standing.

11. p-Toluenesulfonyl chloride is recrystallized from chloroform prior to use.

12. Column chromatography is performed using 30 g of silica per gram of residue and CHCl3/EtOAc (1:1) as the eluent. The resulting clear solution is concentrated and the product crystallizes slowly upon standing.

13. The N-tosylamino derivative **3** has the following spectral data: ^{1}H NMR (300 MHz, CDCl$_3$) (~4:1 mixture of rotamers) δ: 1.05-1.27 (m, 2 H), 1.35-1.54 (m, 1 H), 1.59-1.78 (m, 4 H), 1.80-2.34 (m, 5 H), 2.39 (s, 3 H), 2.61 (dt, 1 H, J = 11.5, 3.2), 3.05 (m, 1 H), 3.49 (m, 1 H), 3.71 (s, 0.6 H, minor rotamer), 3.78 (s, 2.4 H, major rotamer), 3.89 (m, 1 H), 4.48 (dd, 0.8 H, J = 8.5, 4.4, major rotamer), 4.53 (d, 0.2 H, J = 8, exchanges with D$_2$O, minor rotamer), 4.80 (dd, 0.2 H, J = 8.5, 3.0, minor rotamer), 5.41 (d, 0.8 H, J = 2, exchanges with D$_2$O, major rotamer), 7.25 (d, 2 H, J = 8, overlapping minor and major rotamers), 7.69 (d, 0.4 H, J = 8 minor rotamer), 7.74 (d, 1.6 H, J = 8, major rotamer); IR (KBr) cm^{-1}: 3260, 1740, 1611, 1417; chemical ionization mass spectrum, m/z (relative intensity) M$^+$ + 1 (100). Anal. Calcd for C$_{20}$H$_{28}$N$_2$O$_5$S: C, 58.82; H, 6.90. Found: C, 58.92; H, 6.92.

14. The submitters obtained yields of **4** as high as 92% after recrystallization in some runs.

15. The carboxylic acid **4** has the following spectral data: ^{1}H NMR (300 MHz, CDCl$_3$) δ: 1.08-1.30 (m, 4 H), 1.48 (dd, 1 H, J = 24.5, 11.8), 1.56-1.68 (m, 3 H), 1.88-2.00 (m, 2 H), 2.28 (dt, 1 H, J = 11.1, 3.7), 2.38 (m, 4 H), 3.34 (m, 1 H), 5.27 (d, 1 H, J = 8, exchanges with D$_2$O), 7.26 (d, 2 H, J = 8), 7.70 (s (br), 1H, exchanges with D$_2$O), 7.73 (d, 2 H, J = 8); IR (KBr) cm^{-1}: 3310, 1674, 1152; chemical ionization mass spectrum, m/z (relative intensity) 298 (M$^+$ + 1, 30), 280 (100). Anal. Calcd for C$_{14}$H$_{19}$NO$_4$S: C, 56.56; H, 6.44. Found: C, 56.60; H, 6.42.

Waste Disposal Information

All toxic materials were disposed of in accordance with "Prudent Practices for Disposal of Chemicals from Laboratories"; National Academic Press; Washington, DC, 1983.

3. Discussion

Alkali metal in ammonia reductions of pyrrolobenzodiazepine-5,11-diones give trans-2-aminocyclohexanecarboxylic acid derivatives (e.g., **4**) in enantiomerically pure form.[2,3] A method for preparation of cis-2-aminocyclohexanecarboxylic acids related to **4** is based on the enantioselective hydrolysis of symmetrical diesters with pig liver esterase.[4] cis-2-Aminocyclohexane derivatives have been used for syntheses of aminocyclitol antibiotics.[4,5] 6-Alkyl-cis-2-aminocyclohexanecarboxylic acids can be prepared by alkali metal in ammonia reduction of pyrrolobenzodiazepine-5,11-diones followed by olefin hydrogenation; the cis-decahydroquinoline alkaloid (+)-pumiliotoxin C has been prepared by this methodology.[2]

The preparation of **4** described here can be modified to provide a range of substitution patterns as shown in the Table. Some of the derivatives have been used for enantioselective syntheses of poison frog alkaloids that possess the trans-decahydroquinoline ring system.[3,6] The 6-alkyl substituted pyrrolobenzodiazepine-5,11-diones shown in the Table were prepared by metallation of the 6-methylpyrrolobenzodiazepine-5,11-dione with 2 equiv of butyllithium followed by addition of the appropriate alkylation reagent.[3,6] The highly stereoselective alkali metal in ammonia reductions of pyrrolobenzodiazepine-5,11-diones are a special application of a quite general method for enantioselective synthesis of chiral cyclohexanes from benzoic acid derivatives.[7]

1. Department of Chemistry, Rensselaer Polytechnic Institute, Troy, New York 12180-3590.

2. Schultz, A. G.; McCloskey, P. J.; Court, J. J. *J. Am. Chem. Soc.* **1987**, *109*, 6493.

3. McCloskey, P. J.; Schultz, A. G. *J. Org. Chem.* **1988**, *53*, 1380.

4. Kamiyama, K.; Kobayashi, S.; Ohno, M. *Chem. Lett.* **1987**, 29.

5. Kobayashi, S.; Kamiyama, K.; Iimori, T.; Ohno, M. *Tetrahedron Lett.* **1984**, *25*, 2557.

6. Daly, J. W.; Nishizawa, Y.; Padgett, W. L.; Tokuyama, T.; McCloskey, P. J.; Waykole, L.; Schultz, A. G.; Aronstam, R. S. *Neurochem. Res.* **1991**, *16*, 1207.

7. Schultz, A. G. *Acc. Chem. Res.* **1990**, *23,* 207.

TABLE
CHIRAL CYCLOHEXANE DERIVATIVES

Substrate	Product	Yield (%)
Me	Me	91[a]
$CH_3CH_2CH_2$	$CH_3CH_2CH_2$	83[b]
$CH_3(CH_2)_3CH_2$	$CH_3(CH_2)_3CH_2$	90[c]
$HO(CH_2)_2CH_2$	$HO(CH_2)_2CH_2$	92[d]
$HO_2CCH_2CH_2$	$MeO_2CCH_2CH_2$	52[e,f]
$CH_2=CH(CH_2)_2CH_2$	$CH_2=CH(CH_2)_2CH_2$	90[c]
Me	Me	80[a]

[a]See ref. 2. [b]See ref. 3 [c]See ref. 4. [d]Unpublished work of C. Alva. [e]Unpublished work of L. Waykole. [f]The alkali metal in ammonia reduction product was treated with diazomethane.

182

Chemical Abstracts Nomenclature (Collective Index Number);

(Registry Number)

(5aS,9aS,11aS)-Perhydro-5H-pyrrolo[2,1-c][1,4]benzodiazepine-5,11-dione: 1H-
Pyrrolo[2,1-c][1,4]benzodiazepine-5,11(5aH,11aH)-dione, octahydro-,
[5aS-(5aa,9ab,11ab)]- (12); (110419-86-8)

(S)-(+)-2,3-Dihydro-1H-pyrrolo[2,1-c][1,4]benzodiazepine-5,11(10H,11aH)dione: 5H-
Pyrrolo[2,1-c][1,4]benzodiazepine-5,11(10H)-dione, 1,2,3,11a-tetrahydro-, L- (8); 1H-
Pyrrolo[2,1-c][1,4]benzodiazepine-5,11(10H,11aH)-dione, 2,3-dihydro-, (S)- (9);
(18877-34-4)

tert-Butyl alcohol (8); 2-Propanol, 2-methyl- (9); (75-65-0)

Ammonia (8,9); (7664-41-7)

Potassium (8,9); (7440-09-7)

(1S,2S)-2-(N-Tosylamino)-1-[((2S)-2-carbomethoxypyrrolidinyl)carbonyl]cyclohexane:
L-Proline, 1-[[2-[[(4-methylphenyl)sulfonyl]amino]cyclohexyl]carbonyl]-, methyl ester,
(1S-trans)- (12); (110419-91-5)

Triethylamine (8); Ethanamine, N,N-diethyl- (9); (121-44-8)

p-Toluensulfonyl chloride (8); Benzenesulfonyl chloride, 4-methyl- (9); (98-59-9)

(1S,2S)-2-(N-Tosylamino)cyclohexanecarboxylic acid: Cyclohexanecarboxylic acid,
2-[[(4-methylphenyl)sulfonyl]amino]-, (1S-trans)- (12); (110456-11-6)

DIETHYL (2S,3R)-2-(N-tert-BUTOXYCARBONYL)AMINO-3-HYDROXYSUCCINATE

Submitted by Seiki Saito, Kanji Komada, and Toshio Moriwake.[1]
Checked by Martin Fox and Larry E. Overman.

1. Procedure

Caution! Parts A and C of this procedure should be carried out in a well-ventilated hood since toxic hydrogen bromide and azidotrimethylsilane are handled and toxic hydrogen azide is liberated during the course of the reaction.

A. *Diethyl (2S,3S)-2-bromo-3-hydroxysuccinate* (**2**). In an oven-dried, 300-mL, round-bottomed flask containing a Teflon-coated stirring bar is placed 26.0 g (0.126 mol) of diethyl L-tartrate (Note 1). A pressure-equalizing dropping funnel containing 100 mL (0.504 mol) of 30% hydrobromic acid (HBr) in acetic acid (Note 2) is mounted on the flask and the top of the funnel is fitted with a nitrogen inlet vented through an oil bubbler. The system is placed under a nitrogen atmosphere, magnetic stirring is initiated, and contents of the flask are cooled in an ice-water bath. The contents of the dropping funnel are added to the cooled tartrate during 30 min and the yellow mixture is stirred for an additional 15 min after completion of the addition. The cooling bath is removed and the reaction mixture is allowed to reach 25°C and stir for an additional 10 hr or until disappearance of the tartrate as judged by TLC analysis (Note 3). The light brown reaction mixture is poured into 500 g of ice, and the resulting mixture is transferred to a 1-L separatory funnel and extracted four times with 80 mL of ether. The combined ether extracts are washed successively three times with 60 mL of water, and then 100 mL of brine, dried over magnesium sulfate, filtered, and concentrated under reduced pressure using a rotary evaporator to give a pale yellow oil.

A single-necked, 300-mL, round-bottomed flask, equipped with a Teflon-coated stirring bar and an efficient reflux condenser bearing a drying tube packed with blue indicator silica gel beads, is charged with the yellow oil obtained above and 140 mL of ethanol (Note 4) to which is cautiously added 4 mL of acetyl chloride with stirring (Note 5). The mixture is heated under gentle reflux for 7 hr, then cooled to room temperature, the condenser is removed, and the mixture is concentrated under

185

reduced pressure at 50-60°C using a rotary evaporator to give a yellow oil. The crude product is transferred to a silica gel column (Note 6) with the aid of a small amount of hexane-ethyl acetate (4:1) and eluted with the same mixed solvent in 60-mL fractions. Fractions 5 to 15 (Note 7) are combined and concentrated on a rotary evaporator to give a colorless oil, which upon vacuum distillation, affords 24.36-25.73 g (72 - 76%) of 99% pure (glc) diethyl (2S,3S)-2-bromo-3-hydroxysuccinate, bp 93-94°C (0.2 mm) [lit.[4] bp 123-125°C (0.6 mm)], $[\alpha]_D^{25.6}$ -16.8° (EtOH, c 5.70), $[\alpha]_D^{27}$ -35.72° (neat) [lit.[4] $[\alpha]_D^{21}$ -28.9° (neat)] (Notes 8-9).

   B. *Diethyl (2R,3R)-2,3-epoxysuccinate* (**3**). An oven-dried, 300-mL, round-bottomed flask, equipped with a Teflon-coated stirring bar and a rubber septum through which is inserted a large bore, needle-tipped nitrogen line vented through an oil bubbler, is charged with 29.7 g (0.110 mol) of diethyl (2S,3R)-2-bromo-3-hydroxy succinate and 80 mL of dry ethanol (Note 10). In a separate, oven-dried, 250-mL, round-bottomed flask, capped with a rubber septum and vented though a bubbler as described above, a solution of sodium ethoxide in ethanol is prepared from 3.05 g of sodium (0.132 mol) and 120 mL of dry ethanol (Note 10). (*Caution: this operation should be conducted in a well-ventilated hood. Sodium is a highly reactive metal, avoid exposure to moisture. Hydrogen gas, which is highly flammable and forms explosive mixtures with air, is rapidly evolved. External cooling may be necessary.*) The two flasks are placed under a nitrogen atmosphere, and connected by means of a long double needle-tipped cannula. Magnetic stirring is initiated, and the ethanolic solution of the succinate **2** is cooled in an ice-water bath for 15 min followed by dropwise addition of the solution of sodium ethoxide in ethanol which is transferred via the cannula under a very slight positive pressure of nitrogen (Note 11). Complete transfer of the base should require 2 hr (Note 12). The reaction mixture is stirred for an additional 20 min at 0-10°C (bath temperature) and then quenched by the addition of 1.43 mL (0.025 mol) of acetic acid. The reaction mixture is diluted with 900 mL of water, transferred to a 2-L separatory funnel, and extracted four times with 100 mL of dichloromethane. The combined organic phases are washed with 100 mL of brine, dried over sodium sulfate, filtered, and concentrated under reduced pressure using a rotary evaporator to give a colorless oil. Vacuum distillation of this crude product affords 17.7 - 18.7 g (85 - 90%) of diethyl (2R,3R)-2,3-epoxysuccinate (**3**), bp 97-98°C (0.9 mm) [Lit.[4] bp 100-104°C (4 mm)], $[\alpha]_D^{25}$ -110° (EtOH, c 5.09), $[\alpha]_D^{28}$ -115° (neat) [Lit.[4] $[\alpha]_D^{21.5}$ -88.47° (ether, c 1.030)] which is homogeneous by glc analysis (Note 9, and 13-15).

C. *Diethyl (2S,3R)-2-azido-3-hydroxysuccinate* (**4**). An oven-dried, 200-mL, round-bottomed flask, containing a Teflon-coated stirring bar, is capped with a septum while hot, and purged with nitrogen until cool via insertion through the septum of a needle-tipped nitrogen line (vented through an oil bubbler) and an open needle. The open needle is removed and the flask is charged with a solution of 3.45 g (0.0282 mol) of 4-(dimethylamino)pyridine (DMAP) in 17 mL of N,N-dimethylformamide (DMF) (Notes 16 and 17) and 1.93 mL (0.0329 mol) of dry ethanol (Note 11) through the septum via syringe. Stirring is initiated and the mixture is cooled in an ice–water bath. To this cold solution is added 17.1 mL (0.122 mol) of azidotrimethylsilane (Note 18) dropwise at 0°C by means of a syringe (Note 19). The heterogeneous mixture is stirred for 15 min at room temperature (25°C) and then a solution of 17.7 g (0.0940 mol) of diethyl (2R,3R)-2,3-epoxysuccinate (**3**) in 50 mL of chloroform (Note 20), prepared in a septum-capped, oven-dried, 100-mL, round-bottomed flask, is rapidly transferred via cannula with the aid of a positive nitrogen pressure as above (Part B). The mixture is stirred at 25°C until epoxide **3** can no longer be detected upon TLC analysis of the reaction mixture (~ 40 hr). The mixture is then diluted with 300 mL of water, transferred to a 1-L separatory funnel, and the organic phase separated. The aqueous phase is extracted four times with 60 mL of a hexane-ether mixture (1:1 v/v). The combined organic phases are added to 20 mL of a magnetically stirred 2.3 N solution of hydrogen chloride in ethanol (Note 21), followed by dilution of the resulting mixture with an additional 50 mL of ethanol and continued stirring at room temperature for 30 min (Note 22). After transfer to a separatory funnel, the solution is washed successively with four 50-mL portions of water, 50 mL of saturated aqueous sodium bicarbonate solution, and 50 mL of saturated aqueous sodium chloride solution, dried over sodium sulfate, and concentrated under reduced pressure using a rotary evaporator followed by exposure to high vacuum affording 18.6 g (86%) of 96-98% pure diethyl (2S,3R)-2-azido-3-hydroxysuccinate (4) as a pale yellow oil, $[\alpha]_D^{20}$ +31.8° (EtOH, *c* 18.1), as judged by NMR analysis (Notes 23 and 24). This material contains 2-4% of the inseparable (2R,3R)-diastereomer (Note 24).

D. *Diethyl (2S,3R)-2-(N-tert-butoxycarbonyl)amino-3-hydroxysuccinate* (5). A 500-mL, single-necked, round-bottomed flask, equipped with a Teflon-coated stirring bar, is charged with a suspension of 0.91 g of 10% palladium on carbon catalyst (Note 25) in 100 mL of ethyl acetate (Note 26). The flask is connected to a normal pressure hydrogenation apparatus (Note 27) and the catalyst is saturated with hydrogen. After removal of the hydrogen, a solution of 18.2 g (0.0785 mol) of 4 and 20.6 g (0.0942 mol) di-tert-butyl dicarbonate (Note 28) in 80 mL of ethyl acetate (Note 26) is added to the suspension of catalyst, a hydrogen atmosphere reestablished, and the suspension is stirred at room temperature under a slight positive pressure of hydrogen for 4-6 hr (Note 29). The suspension is filtered through a Celite pad, and the pad is rinsed with several portions of ethyl acetate. The combined ethyl acetate solutions are concentrated on a rotary evaporator and finally under high vacuum to give a pale yellow oil that is initially purified by means of a column packed with silica gel (100 g) using hexane-ethyl acetate (6:1) as eluent (Note 30). Fractions containing the product are combined and concentrated on a rotary evaporator to give 23.3 g of crude 5 as a colorless oil. The oily crude 5 is dissolved in 70 mL of hexane-ether (3:1), and the solution is cooled to -30°C, seeded, and kept overnight at that temperature (freezer) to allow crystallization (Note 31). The mother liquor is siphoned out while the mixture is kept at -30°C (dry ice-acetone bath). The crystals are washed with several portions of hexane-ether (3:1) at -30°C, then dried under high vacuum to provide 12.2 - 12.7 g of diastereomerically and enantiomerically pure diethyl (2S,3R)-2-(N-tert-butoxy-carbonyl)amino-3-hydroxysuccinate (5) as colorless prisms, mp 33-34°C; $[\alpha]_D^{24}$ +13.1° (EtOH, c 6.32); $[\alpha]_D^{24}$ +28.7° (CHCl$_3$, c 6.05) (Note 32). The combined mother liquor and the hexane-ether (3:1) washings are concentrated on a rotary evaporator to give a colorless oil, which upon crystallization as above provides an additional 2.7-3.8 g of product 5 (Note 33). The combined yield of crystalline 5 is 15.9-16.5 g (66 - 73%).

2. Notes

1. Diethyl L-tartrate was usually purchased from Aldrich Chemical Co., Ltd. The diester was also prepared from L-tartaric acid by the procedure of Kocienski[2] (triethyl orthoformate–ethanol–acetyl chloride) or Seebach[3] [ethanol–acidic ion-exchange resin (Lewatit 3333)].

2. A solution of HBr (30%) in acetic acid (d = 1.31 g/mL) was purchased from Kishida Chemical Co., Ltd. (Japan). When 3 equiv of HBr rather than 4 equiv was employed, the yield of 2 dropped by more than 40%.

3. Using Merck precoated silica gel plates (0.25-mm thickness) with elution by hexane-EtOAc (2:1), after this time period, the starting tartrate ester (R_f = 0.14) was absent. Three spots with R_f values of 0.53 (the largest), 0.36 (half of the largest), and 0.21 (trace) were observed upon visualization. See discussion for the structure of these products.

4. Commercial reagent grade ethanol was used as received.

5. Acetyl chloride was purchased from Kishida Chemical Co. Ltd. (Japan), and used as received.

6. A column (350 mm x 50 mm) packed with Merck silica gel 60 (230-400 mesh ASTM) was used. Elution was facilitated by the action of an air pump for tropical fish tanks available from supermarkets. After use, the column was washed successively with methanol (500 mL), ethyl acetate (400 mL), and hexane (400 mL). Such a washing operation reactivates the column enough for reuse to purify the epoxide (**3**), if necessary (see Note 15). Thus the submitters can employ the same column repeatedly for the purification of **2** and **3** (over 10 times). The checkers recommend purification by flash chromatography using a 350-mm x 50-mm column.

7. Fractions 17 to 27, 25-27 eluted with hexane-ethyl acetate (1:2), afforded recovered diethyl L-tartrate (3.1 g).

8. The spectral data for **2** are as follows: ^{1}H NMR (500 MHz, CDCl$_3$) δ: 1.31 (t, 6 H, J = 7.1), 3.42 (d, 1 H, J = 7.5), 4.21-4.35 (m, 4 H), 4.66 (dd, 1 H, J = 7.5, 4.1), 4.71 (d, 1 H, J = 4.1); ^{13}C NMR (50 MHz, CDCl$_3$) δ :13.7, 13.8, 47.4, 62.3, 62.6, 72.3, 166.5, 170.1; IR (film) cm^{-1}: 3480, 2909, 2873, 1739, 1371, 1302, 1286, 1220, 1161, 1113, 1024.

9. A Simadzu GC–8A gas chromatograph equipped with a flame ionization detector and ULBON HR-101 capillary column (23 m x 0.25 mm) was used.

10. Ethanol was freshly distilled from magnesium ethoxide.

11. If it proves difficult to control the rate of addition by adjusting the nitrogen pressure, a static positive pressure of nitrogen can be employed and the flask containing the sodium ethoxide solution can be moved up or down slightly to regulate the rate of addition.

12. More rapid addition of the base results in a lower yield of **3**.

13. The spectral data for **3** are as follows: ^{1}H NMR (300 MHz, CDCl$_3$) δ: 1.31 (t, 6 H, J = 7.1), 3.66 (s, 2 H), 4.21-4.32 (m, 4 H); ^{13}C NMR (50 MHz, CDCl$_3$) δ: 13.9, 51.9, 62.1, 166.6; IR (film) cm^{-1}: 2988, 1751, 1372, 1330, 1200, 1029.

14. Epoxide **3** should be free from **2** because **2**, if subjected to the next step (**3** → **4**), leads to the (2R,3R)-diastereoisomer via an S_N2 displacement process. If **3** is contaminated with **2**, the distillate can be purified using the same column as above (Note 6), employing hexane–ether (5:1) as eluent and 60-mL fractions. In the checkers' hands a trace of **2** was always present. Therefore, the checkers recommend that the crude product be purified by flash chromatography (as in step A) to remove residual **2** prior to distillation.

15. The literature values[4] for the antipode (95% pure by GLC analysis) are as follows: bp 98-99°C (3 mm) and $[\alpha]_D^{23}$ +105.49° (ether, c 1.413).

16. DMAP was purchased from Nacalai Tesque, Inc. (Japan) and used as received.

17. N,N-Dimethylformamide was distilled from calcium hydride.

18. Azidotrimethylsilane was purchased from Aldrich Chemical Co., Ltd. (95% pure) and used as received.

19. The reaction of azidotrimethylsilane and ethanol, which gives hydrogen azide and ethoxytrimethylsilane, is exothermic and the hydrogen azide immediately forms a salt with 4-(dimethylamino)pyridine, a part of which cannot be dissolved in N,N-dimethylformamide and precipitates.

20. Commercial, reagent grade chloroform was used as received.

21. The solution was prepared by carefully mixing 3.28 mL acetyl chloride with sufficient dry ethanol to reach a final volume of 20 mL.

22. This operation transforms any diethyl (2S,3R)-2-azido-3-(trimethylsiloxy)-succinate present to the desired diethyl (2S,3R)-2-azido-3-hydroxysuccinate.

23. The optical rotation of **4** is highly dependent both on solvent and sample concentration. For example, $[\alpha]_D$ values observed in chloroform vary as follows: +1.43° (c 3.25) , +3.69° (c 6.61), and +5.68° (c 16.7).

24. The spectral data for **4** are as follows: ^1H NMR (500 MHz, CDCl$_3$) δ: 1.30 (t, 3 H, J = 7.1), 1.31 (t, 3 H, J = 7.1), 3.38 (d, 1 H, J = 5.4), 4.20-4.34 (m, 5 H), 4.63 (dd, 1 H, J = 5.4, 2.7); ^{13}C NMR (50 MHz, CDCl$_3$) δ: 13.90, 13.94, 62.3, 62.6, 64.3, 72.0, 166.9, 170.7; IR (film) cm^{-1}: 3480, 2989, 2943, 2910, 2877, 2120, 1751, 1724, 1209, 1114, 1028. The signal corresponding to C$_3$ proton of the (2R,3R) diastereoisomer appears at δ 4.73 (dd, J = 5.6, 2.4).

25. 10% Palladium on carbon catalyst was purchased from Ishizu Seiyaku Co., Ltd. (Japan) and employed as received.

26. Commercial reagent-grade ethyl acetate was used as received. Ethyl acetate is the best choice of solvent with regard to reaction time. Alcohols such as methanol or ethanol should be avoided because di-tert-butyl dicarbonate is decomposed by these solvents.

27. An MRK catalytic hydrogenation apparatus (Mitamura Riken Co., Japan) was used.

28. Di-tert-butyl dicarbonate (97% pure) was purchased from Wako Pure Chemical Industries, Ltd. (Japan) and used as received.

29. The time required for completion of the reaction depends on the activity of the palladium catalyst employed.

30. This solvent system is convenient to separate less polar contaminants such as tert-butyl acetate or unchanged di-tert-butyl dicarbonate.

31. The seeds required for crystallization are available by cooling a small portion (1 mL) of the hexane-ether solution to -78°C and scratching the highly viscous gum that separates with a glass rod. The rate of crystallization is very slow and additional ether is sometimes required when an oil phase appears on cooling instead of development of crystals.

32. When commercial chloroform (stabilized with 0.3-1% of ethanol) was used as received for optical rotation measurements, the submitters obtained $[\alpha]_D^{24}$ +19.3° (c

6.05). However, chloroform passed through a silica gel column prior to the measurement gave a much higher value as indicated. The optical rotation of **5** observed with commercial ethanol was ($[\alpha]_D^{24}$ +19.3° (EtOH, c 6.32)), while that determined using absolute EtOH [distilled from Mg(OEt)$_2$] led to a much lower value $[\alpha]_D^{24}$ +7.19° (EtOH, c 6.32). The spectral properties for pure **5** are as follows: ^{1}H NMR (CDCl$_3$, 300 MHz) δ: 1.25 (t, 3 H, J = 7.1), 1.32 (t, 3 H, J = 7.1), 1.46 (s, 9 H), 3.47 (s (br), 1 H), 4.14-4.34 (m, 4H), 4.50 (s (br), 1 H), 4.83 (dd, 1 H, J = 7.9, 1.3), 5.51 (d, 1 H, J = 7.9); ^{13}C NMR (CDCl$_3$, 50 MHz) δ: 13.9, 14.0, 28.2, 56.9, 61.8, 62.1, 72.0, 80.3, 155.5, 168.5, 171.5; IR (film) cm[-1]: 3450 (br), 2983, 1720, 1507, 1369, 1241, 1218, 1188, 1117, 1059, 1029.

33. After isolation of the second crop, the recovered oily product mixture should be purified, if required, by column chromatography (100 g of silica gel/g oil) using hexane–ethyl acetate (8:1). The rate of crystallization for the third crop becomes much slower, requiring 2 to 3 days at -30°C; cooling at -78°C results in oil separation. Careful chromatography on silica gel of the oily residue remaining after concentration of the mother liquors of the third crystallization affords the pure 2R,3R diastereomer. Spectral data for diethyl (2R,3R)-2-(N-tert-butoxycarbonyl)amino-3-hydroxy–succinate are as follows: ^{1}H NMR (CDCl$_3$, 500 MHz) δ: 1.30 (t, 3 H, J = 6.9), 1.31 (t, 3 H, J = 7.0), 1.41 (s, 9 H), 3.45 (d, 1 H, J = 5.0), 4.22-4.30 (m, 4 H), 4.68 (d (br), 1 H, J = 3.3), 4.78 (d (br), 1 H, J = 8.3), 5.26 (d (br), 1H, J = 9.3); ^{13}C NMR (CDCl$_3$, 50 MHz) δ: 13.9, 14.0, 28.0, 56.0, 61.9, 62.4, 71.1, 80.0, 155.2, 169.4, 171.9.

Waste Disposal Information

All toxic materials were disposed of in accordance with "Prudent Practices for Disposal of Chemicals from Laboratories"; National Academic Press; Washington, DC, 1983.

3. Discussion

Optically active β-hydroxy-α-amino acids are important not only as constituents of biologically active peptides,[5] but also as precursors to β-lactam antibiotics.[6] Various synthetic approaches to compounds of this class have been reported involving the traditional resolution of racemic compounds,[7] asymmetric syntheses,[8] or the derivatization of appropriate chiral templates.[9] The procedure described here is a practical preparation of diethyl erythro-3-hydroxy-N-(tert-butoxycarbonyl)-L-aspartate (5) in high optical purity, amenable to large scale preparation. The synthetic route involves the transformation of diethyl L-tartrate to diethyl (2S,3S)-2-bromo-3-hydroxy succinate (2),[4] preparation of diethyl (2R,3R)-2,3-epoxysuccinate (3),[4] nucleophilic cleavage of the epoxide by azide to give diethyl (2S,3R)-2-azido-3-hydroxysuccinate (4),[9f] and the single-pot transformation of the azido group into the N-(tert-butoxycarbonyl)amino group[10] to give 5 in high enantiomeric and diastereomeric purity.

Scheme I

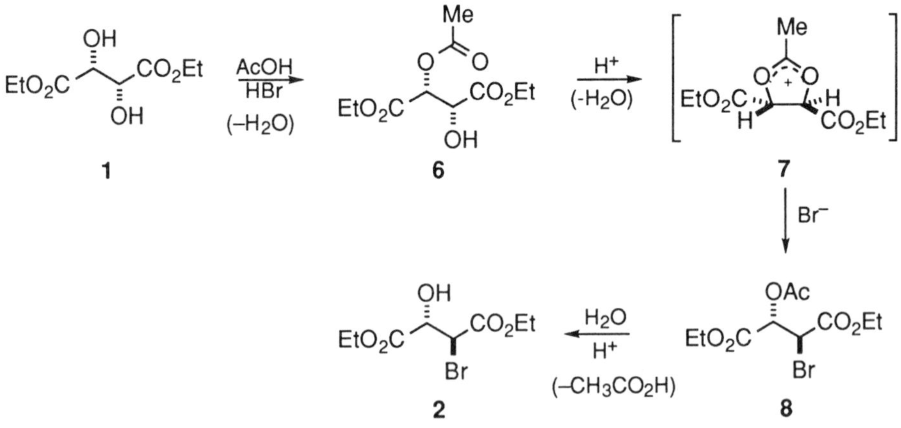

194

The first and second steps are examples of a standard method for the stereospecific synthesis of epoxides via vicinal acetoxy bromides from vicinal diols.[11] Based upon kinetic studies, the stereochemical outcome, and the observation of an intermediate carbocation by NMR spectroscopy,[11a] the mechanism of the first step has been rationalized to involve monoacetylation of the diol, cyclization to a 1,3-dioxolan-2-ylium ion, and capture of this intermediate by bromide ion. If this mechanism is valid for the present case, the reaction can be illustrated as shown in Scheme I. The experimental findings (Note 3) correspond reasonably with this mechanistic scheme. Reaction of 1 with 30% HBr in acetic acid proceeded, as shown by both TLC and isolation experiments, with the initial formation of 8 (R_f = 0.53) followed by the gradual increase in the formation of 2 (R_f = 0.36) because of increase in the content of water liberated in the medium. At complete consumption of 1, an additional minor product 6 (R_f = 0.21) was detected by TLC. The presence of 6 may result from competition by the increasing concentration of water as the reaction progresses for 7 relative to bromide. Thus, although 1 had been completely consumed as observed by TLC, the subsequent transesterification process (ethanol/HCl) gives rise to a mixture of 2 and a small amount of 1 (derived from 6) which are easily separated by column chromatography on silica gel.

The optimal base for closure to epoxide 2 is sodium ethoxide in ethanol, as potassium hydroxide in ethanol, sodium or potassium carbonate in ethanol, or benzyltrimethylammonium hydroxide in ethanol resulted in the formation of a complex mixture of decomposition products arising mainly from deprotonation at carbon.

For the azide cleavage of chiral 2,3-epoxy diester 3, the submitter's earlier procedure[9f] was modified to reduce the amount of azide reagent (azidotrimethylsilane, TMSN$_3$) and lower the reaction temperature. This effort, coupled with the submitter's recent finding that such a reaction can be accelerated in the presence of amines,[8g,9i] led to the use of the system described above. The reaction may proceed by way of the pathway shown in Scheme II. The reactive azide species could be $DMAPH^+N_3^-$, formed from DMAP and HN_3 which are, in turn, initially generated by the exothermic reaction of $TMSN_3$ and ethanol, or later catalytically from reaction of $TMSN_3$ with the desired product **4**. The reaction occurs with these reagents at a practical rate at 25°C, in sharp contrast to the previous more vigorous conditions (2 equiv or more of TMSN3-

Scheme II

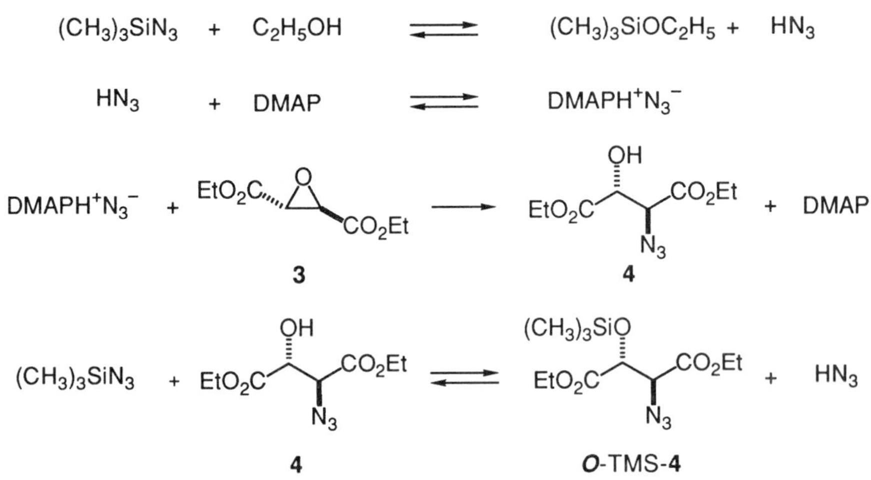

$(CH_3)_3SiN_3$ + C_2H_5OH ⇌ $(CH_3)_3SiOC_2H_5$ + HN_3

HN_3 + DMAP ⇌ $DMAPH^+N_3^-$

$DMAPH^+N_3^-$ + **3** → **4** + DMAP

$(CH_3)_3SiN_3$ + **4** ⇌ **O-TMS-4** + HN_3

CH$_3$OH (1:1) at 60°C in DMF).[9f] In addition, this reaction does not require any special precautions other than use of a well-ventilated hood since the generation a high concentration of the hazardous free HN_3 is avoided.[8g,9i]. The initial product is a mixture of **4** and its O-silylated derivative (O-TMS-4). The latter is converted to **4** upon treatment of the mixture with HCl in ethanol affording **4** in over 85% yield.

The final one-pot, two-stage conversion of azide to an amino group and then to the related tert-butoxycarbonyl derivative was effected as before.[10] Catalytic hydrogenolysis of **4** over 10% Pd-carbon in ethyl acetate in the presence of di-tert-butyl dicarbonate for 4 hr at room temperature provides a crude product consisting of **5** and its (2R,3R)-isomer (in the same ratio as that observed for **4** and its (2R,3R)-isomer) that can be separated by fractional crystallization from a hexane-ether mixed solvent at -30°C.

This procedure has been used by Ohfune[12] to convert N-(benzyloxycarbonyl)amino groups to N-(tert-butoxycarbonyl)amino groups in one pot, the Cbz to Boc switching protocol. This procedure is efficient because no tedious isolation of the intermediate amine is required. The following examples illustrate the advantages of this one-pot procedure. In addition, the amount of palladium catalyst employed controls the chemoselectivity of the transformation. For instance, employing 1-2 wt% Pd/g of substrate, the 1,2-diazide **9** can be changed to the corresponding, protected 1,2-diamine derivative **10**, while the O-benzyl group is kept unchanged.[9i] On the other hand, employing 30 wt% Pd/g of substrate, both the azido and O-benzyl groups of **11** can be transformed to N-(tert-butoxycarbonyl)amino and hydroxyl groups, (**12**) respectively (Scheme III).[13]

Scheme III

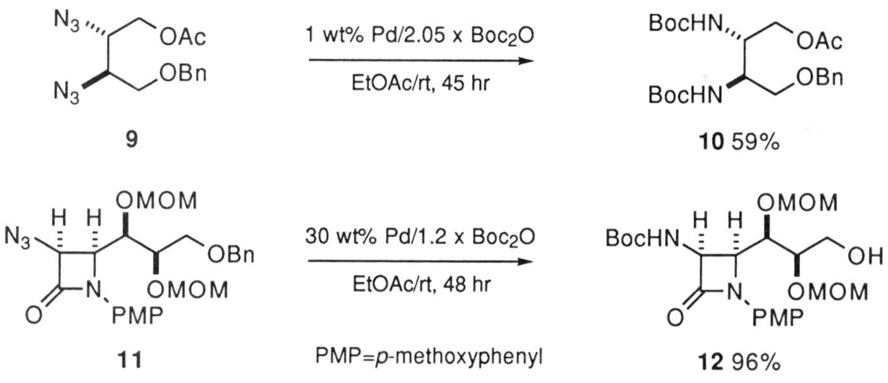

PMP=*p*-methoxyphenyl

A recent report by Evans[14] of azide reduction by $SnCl_2$ in aqueous dioxane and in situ amine protection by addition of Boc_2O has demonstrated that other methods of reduction of azides may be feasible if catalytic reduction of the azide is precluded because of the presence of protective groups sensitive to hydrogenolysis.

1. Department of Applied Chemistry, Faculty of Engineering, Okayama University, Okayama 700, Japan.

2. Kocienski, P.; Street, S. D. A. *Synth. Commun.* **1984**, *14*, 1087.

3. Seebach, D.; Kalinowski, H.–O.; Bastani, B.; Crass, G.; Daum, H.; Dörr, H.; DuPreez, N.P.; Ehrig, V.; Langer, W.; Nüssler, C.; Oei, H.–A.; Schmidt, M. *Helv. Chim. Acta* **1977**, *60*, 301.

4. Mori, K.; Iwasawa, H. *Tetrahedron* **1980**, *36*, 87.

5. (a) "Amino Acids, Peptides and Proteins," Specialist Periodical Report, Chemical Society: London, 1968-1983; Vol. 1-16; (b) Rich, D. H.; Sun, C. Q.; Guillaume, D.; Dunlap, B.; Evans, D. A.; Weber, A. E. *J. Med. Chem.* **1989**, *32*, 1982.

6. (a) Holden, K. G. in "Cephalosporins and Penicillins: Chemistry and Biology," Flynn, E. H. Ed.; Academic Press: New York, 1972, Vol. 2, pp 133-136; (b) Floyd, D. M.; Fritz, A. W.; Pluscec, J.; Weaver, E. R.; Cimarusti, C. M. *J. Org. Chem.* **1982**, *47*, 5160; (c) Labia, R.; Morin, C. *J. Antibiotics* **1984**, *37*, 1103; (d) Miller, M. J. *Acc. Chem. Res.* **1986**, *19*, 49.

7. Liwschitz, Y.; Edilitz-Pfeffermann, Y.; Singerman, A. *J. Chem. Soc. C* **1967**, 2104.

8. (a) Nakatsuka, T.; Miwa, T.; Mukaiyama, T. *Chem. Lett.* **1981**, 279; (b) Chong, J. M.; Sharpless, K. B. *J. Org. Chem.* **1985**, *50*, 1560; (c) Shioiri, T.; Hamada, Y. *Heterocycles* **1988**, *27*, 1035; (d) Evans, D. A.; Weber, A. E. *J. Am. Chem. Soc.* **1986**, *108*, 6757; (e) Evans, D. A.; Sjogren, E. B.; Weber, A. E.; Conn, R. E.

Tetrahedron Lett. **1987**, *28*, 39; (f) Schmidt, U.; Siegel, W. *Tetrahedron Lett.* **1987**, *28*, 2849; (g) Saito, S.; Takahashi, N.; Ishikawa, T.; Moriwake, T. *Tetrahedron Lett.* **1991**, *32*, 667.

9. (a) Saito, S.; Bunya, N.; Inaba, M.; Moriwake, T.; Torii, S. *Tetrahedron Lett.* **1985**, *26*, 5309; (b) Saito, S.; Yamashita, S.; Nishikawa, T.; Yokoyama, Y.; Inaba, M.; Moriwake, T. *Tetrahedron Lett.* **1989**, *30*, 4153; (c) Saito, S.; Nishikawa, T.; Yokoyama, Y.; Moriwake, T. *Tetrahedron Lett.* **1990**, *31*, 221; (d) Saito, S.; Yokoyama, H.; Ishikawa, T.; Niwa, N.; Moriwake, T. *Tetrahedron Lett.* **1991**, *32*, 663.

10. Saito, S.; Nakajima, H.; Inaba, M.; Moriwake, T. *Tetrahedron Lett.* **1989**, *30*, 837.

11. (a) Golding, B. T.; Hall, D. R.; Sakrikar, S. *J. Chem. Soc., Perkin I Trans.* **1973**, 1214; (b) Schmidt, U.; Talbiersky, J.; Bartkowiak, F.; Wild, J. *Angew. Chem., Int. Ed. Engl.* **1980**, *19*, 198.

12. Sakaitani, M.; Hori, K.; Ohfune, Y. *Tetrahedron Lett.* **1988**, *29*, 2983.

13. Saito, S.; Ishikawa, T.; Moriwake, T. *Synlett* **1993**, 139.

14. Evans, D. A.; Evrard, D. A.; Rychnovsky, S. D.; Früh, T.; Whittingham, W. G.; DeVries, K. M. *Tetrahedron Lett.* **1992**, *33*, 1189.

Appendix
Chemical Abstracts Nomenclature (Collective Index Number); (Registry Number)

Diethyl (2S,3S)-2-bromo-3-hydroxysuccinate: Butanedioic acid, 2-bromo-3-hydroxy-, diethyl ester, [S-(R*,R*)]- (11); (80640-14-8)

Diethyl (2R,3R)-tartrate: Tartaric acid, diethyl ester, L-(+)- (8); Butanedioic acid, 2,3-dihydroxy-, [R-(R*,R*)]-, diethyl ester (9); (87-91-2)

Acetyl chloride (8,9); (75-36-5)

Diethyl (2R,3R)-2,3-epoxysuccinate: 2,3-Oxiranedicarboxylic acid, diethyl ester, (2R-trans)- (10); (74243-85-9)

Diethyl (2S,3R)-2-azido-3-hydroxysuccinate: Butanedioic acid, 2-azido-3-hydroxy-, diethyl ester, [S-(R*,S*)] (12); (101924-62-3)

4-Dimethylaminopyridine: Pyridine, 4-(dimethylamino)- (8); 4-Pyridinamine, N,N-dimethyl- (9); (1122-58-3)

N,N-Dimethylformamide (Cancer Suspect Agent): Formamide, N,N-dimethyl- (8,9); (68-12-2)

Azidotrimethylsilane (Highly toxic): Silane, azidotrimethyl- (8,9); (4648-54-8)

Di-tert-butyl dicarbonate: Formic acid, oxydi- di-tert-butyl ester (8); Dicarbonic acid, bis(1,1-dimethylethyl) ester (9); (24424-99-5)

SYNTHESIS OF ENANTIOMERICALLY PURE β-AMINO ACIDS FROM 2-tert-BUTYL-1-CARBOMETHOXY-2,3-DIHYDRO-4(1H)-PYRIMIDINONE: (R)-3-AMINO-3-(p-METHOXYPHENYL)PROPIONIC ACID

(1(2H)-Pyrimidinecarboxylic acid, 2-(1,1-dimethylethyl)-3,4-dihydro-4-oxo-, methyl ester, (R)- or (S)-)

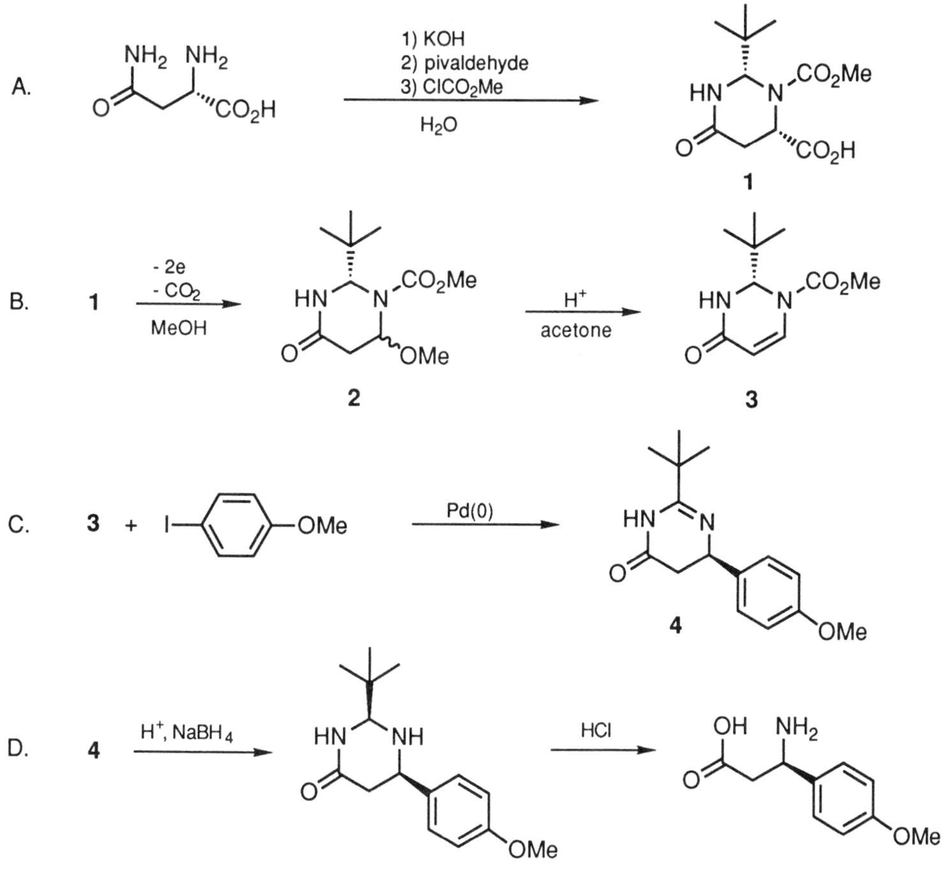

A.

NH_2 NH_2

O CO_2H

1) KOH
2) pivaldehyde
3) $ClCO_2Me$

H_2O

CO_2Me
HN   N
O CO_2H
1

B.   **1**

$-2e$
$-CO_2$

MeOH

CO_2Me
HN   N
O OMe
2

H^+
acetone

CO_2Me
HN   N
O
3

C.   **3** + I—⟨ ⟩—OMe

Pd(0)

HN   N
O
OMe
4

D.   **4**

H^+, $NaBH_4$

HN   NH
O
OMe

HCl

OH NH_2
O
OMe

Submitted by F. J. Lakner, K. S. Chu, G. R. Negrete, and J. P. Konopelski.[1]

Checked by Pradeep B. Madan, George P. Yiannikouros, and David L. Coffen.

1. Procedure

A. (S,S)-2-tert-Butyl-1-carbomethoxy-6-carboxy-2,3,5,6-tetrahydro-4(1H)-pyrimidinone, **1**. An Erlenmeyer flask equipped with a magnetic stirring bar is charged with 21.8 g of potassium hydroxide (KOH, 85% assay, 0.33 mol) and 500 mL of deionized water. A 50-g portion of L-asparagine monohydrate (0.33 mol) is added with stirring, followed by the addition of 43.6 mL of pivalaldehyde (34.4 g, 0.4 mol) with vigorous stirring, after which the mixture becomes homogeneous (Note 1). At the end of 6 hr, the solution is chilled in an ice bath and 28.0 g of sodium bicarbonate (NaHCO$_3$, 0.33 mol) is added, followed by 25.8 mL of methyl chloroformate (0.33 mol) (Note 1). After 1 hr of vigorous stirring at ice temperature, an additional 8.5 g of NaHCO$_3$ (0.11 mol) and 7.5 mL of methyl chloroformate (0.11 mol) are added. The ice bath is removed and the solution is allowed to warm to room temperature for 2 hr. Precipitation of heterocycle **1** is accomplished by slow addition of 125 mL of 10% hydrochloric acid (0.44 mol). The resulting solids are collected by vacuum filtration, rinsed with 225 mL of ice water and dried in a vacuum desiccator yielding 62.1 - 67.1 g (72 - 79%) of heterocycle **1** as a white solid, mp 201°C (dec), [α]$_D$ -108.5° (CH$_3$OH, *c* 2.05) (Note 2).

B. (S)-2-tert-Butyl-1-carbomethoxy-2,3-dihydro-4(1H)-pyrimidinone, **3**. A 1-L, three-necked, round-bottomed flask is equipped with two electrodes, a thermometer, and a magnetic stirring bar (Note 3). The flask is charged with a solution of 60.0 g of heterocycle **1** (0.232 mol) in 620 mL of methanol (Note 4) and 3.2 mL (2.34 g, 23.2 mmol) of triethylamine (Note 5). The temperature is maintained at ≤ 20°C with a circulating water bath, while 0.60 amp is applied for 26 hr (Note 6). After transfer to a suitable vessel, the solvent is evaporated under reduced pressure providing heterocycle **2** as a clear, colorless syrup (Note 7). The residue is dissolved in 425 mL of acetone (Note 4), 20 g of cation exchange resin (Note 8) is added, and the suspension is then heated at reflux for 1 hr. The resin catalyst is collected by filtration and rinsed with an additional 100 mL of acetone, and the filtrate is concentrated under reduced pressure affording heterocycle **3** as a light yellow gum. The crude material is dissolved in 150 mL of ethyl acetate and washed twice with 50-mL portions of saturated aqueous sodium bicarbonate. The organic layer is dried with magnesium sulfate ($MgSO_4$), filtered, and crystallization is allowed to occur by slow evaporation of the solvent. The crystals are collected affording 30-46.8 g (48 - 68% from asparagine) of large, clear, colorless crystals of heterocycle **3**, mp 142-143°C, $[\alpha]_D$ +416.6° (EtOAc, c 1.70) (Note 9).

C. *(R)-2-tert-Butyl-6-(4-methoxyphenyl)-5,6-dihydro-4(1H)-pyrimidinone*, **4**. Compound **3** (5.00 g, 23.6 mmol), 4-iodoanisole (5.52 g, 23.6 mmol), diethylamine (Et_2NH) (2.68 mL, 26.0 mmol), and tetrakis(triphenylphosphine)palladium(0) [$Pd(PPh_3)_4$] (273 mg, 0.24 mmol) are dissolved in 25 mL of dimethylformamide (Note 10). The yellow-colored solution is transferred to a thick-walled Pyrex pressure tube and purged with argon. The tube is then sealed and suspended in a boiling water bath for 48 hr in the dark (Note 11). The tube is cooled in liquid nitrogen before opening (**Caution: pressure has developed**). The contents of the tube are transferred with the aid of 250 mL of ethyl acetate (EtOAc) to a wet packed approx. 40-mm x 50-mm bed of 17.5 g silica gel that is then rinsed with an additional 200 mL of EtOAc. After the solvent is removed under vacuum, the residue is dissolved in a mixture of 80 mL of water, 21.0 mL of 1.0 N hydrochloric acid solution, and 50 mL of ether. The resulting solution is transferred to a 500-mL separatory funnel, washed twice with 100-mL portions of ether, and the ethereal washings are discarded. The aqueous layer is transferred to an Erlenmeyer flask, cooled in an ice bath, and ~11.5 mL of 2 N aqueous sodium hydroxide solution is added in one portion with stirring to bring the solution to a pH of 8-9. The resulting precipitate (Note 12) is collected by

203

suction filtration and air dried to give **4** as a yellowish solid (3.33-4.50 g, 55-75%), mp 118-120°C, $[\alpha]_D$ -20° (CHCl$_3$, *c* 3.3) (Note 13).

    D. (R)-3-Amino-3-(p-methoxyphenyl)propionic acid.[2] A 125-mL Erlenmeyer flask containing a magnetic stirring bar is charged with 20 mL of tetrahydrofuran, 20 mL of 95% ethanol, and heterocycle **4** (2.39 g, 9.2 mmol). The mixture is stirred and cooled to -35°C to -45°C with an acetone-dry ice bath. About 0.6 mL of aqueous 9 N hydrochloric acid is added dropwise until a pH of approximately 7 is obtained as determined by pH paper. A solution of sodium borohydride, prepared by dissolving 0.5 g of sodium borohydride (13.0 mmol) in approximately 1.5 mL of water (minimum amount) containing 1 drop of 30% aqueous sodium hydroxide solution, is added dropwise alternately with 9 N hydrochloric acid to the stirred solution of **4** such that a pH of 6-8 is maintained. During the addition, the bath temperature should be maintained between -35°C and -45°C. After the addition of sodium borohydride solution is complete, the reaction mixture is stirred at -35°C for 1 hr. During this time, the reaction mixture is maintained at a pH of 7 by occasional addition of 9 N hydrochloric acid (about 0.4 mL additional is required). The reaction mixture is then stored at -20°C overnight. After warming to room temperature, the reaction mixture is transferred to a separatory funnel and the pH is raised to 9 by addition of aqueous 40% sodium hydroxide (1.5 mL). After dilution with 30 mL of water, the mixture is extracted three times with 20-mL portions of ether, and the combined ether extracts are washed with 10 mL of saturated sodium chloride. After the organic layer is dried over potassium carbonate, the solution is concentrated under reduced pressure giving 2.01 g (85%) of the acyl aminal as a slightly yellow solid that is used without further purification (Note 14).

    An 0.80-g (3.0 mmol) portion of the crude aminal above is dissolved in 8 mL of 4.5 N hydrochloric acid solution and heated to 100°C in a boiling water bath for 2.5 hr. The clear liquid reaction mixture is transferred to an evaporating dish and left to evaporate in a fume hood. The residue is dissolved in 2 mL of 2 N hydrochloric acid solution and the pH of the solution is adjusted to 7 by slow addition of 30% aqueous sodium hydroxide with swirling, at which time the whole mixture solidifies. The solid mixture is kept at -20°C overnight, followed by addition of 10 mL of water and stirring with a glass rod. The solids are collected by suction affording 0.38 - 0.41 g (63-69%) of the β-amino acid of ~85% purity as a slightly yellow crystalline solid, mp 235°C (dec); $[\alpha]_D$ -4° (1 N HCl, c 1.86) (Note 15).

2. Notes

1. L-Asparagine monohydate, pivalaldehyde, and methyl chloroformate were purchased from the Aldrich Chemical Company, Inc. and used without purification.

2. An extra 8.0 g may precipitate by saturation of the filtrate with sodium chloride and storing the solution in a refrigerator overnight. This material is of significantly inferior chemical and diastereomeric purity. The initial precipitate, however, is usually 92% optically pure as compared to the material obtained by recrystallization from 1:4 ethanol/water, and exhibits constant rotation upon further recrystallization. The rotation for twice recrystallized **1** is $[\alpha]_D$ -115° (CH$_3$OH, c 2.05). However, this purification procedure (recrystallization) is not necessary, since heterocycle **3** is easily crystallized to ≥ 99% ee as the final isolation step. Spectral characteristics for pure **1** are: ^{1}H NMR (400 MHz, D$_2$O/K$_2$CO$_3$, 10 mg/mL) δ: 1.16 (s, 9 H), 2.87-2.96 (m, 2 H), 3.83 (s, 3 H), 4.58 (m, 1 H), 5.33 (s, 1 H); IR (KBr) cm^{-1}: 3283, 2966, 1719, 1631, 1314, 1220, 1090, 779.

3. The flask possesses three vertical side necks of equal height. The middle neck and one side neck were fitted with 1-hole septa pierced by 1-cm diameter cylindrical graphite electrodes (Sargent Welch). The electrodes were inserted 7 cm into the solution, supplying a working electrode surface of 23 cm^2, and were 5 cm

apart. A thermometer was inserted through the third neck with room to vent gases. A power supply (Southwest Technical Products Corp.) is attached to the electrodes with alligator clips. These conditions provide 34-37 V at 0.60 A.

4. Methanol was purchased from Fisher Scientific Company and used without further purification.

5. Triethylamine was purchased from the Aldrich Chemical Company, Inc., distilled, and stored over KOH pellets.

6. This represents the time necessary to pass 2.5 F/mol for the quantity stated at 0.60 A.

7. Methoxylated product **2** consists of diastereomers in a 3:1 ratio, the main proton resonances being ^1H NMR (400 MHz, $CDCl_3$) δ: 0.95 and 1.02 (s, 9 H), 2.6-2.9 (m, 2 H), 3.35 and 3.39 (s, 3 H), 3.77 (s, 3 H), 5.17-5.29 (m, 1 H), 8.07-8.1 (s (br), 1 H).

8. Dowex 50W-X8 cation exchange resin, 200-400 mesh, in the hydrogen form was purchased from J. T. Baker Chemical Company. It was placed in a sintered-glass funnel and washed with three bed-volumes each of 10% hydrochloric acid, water, methanol, and acetone, and then dried under vacuum. The checkers recommend Soxhlet extraction of the resin for complete recovery of product **3**.

9. Product **3** has the following spectral properties: ^1H NMR (400 MHz, $CDCl_3$) δ: 0.95 (s, 9 H), 3.83 (s, 3 H), 5.15-5.45 (m, 2 H), 7.2-7.5 (m, 1 H), 8.0-8.5 (m, 1 H); ^{13}C NMR (100 MHz, $CDCl_3$) δ: 25.5, 40.7, 54.0, 72.0, 104.8, 137.5, 153.3, 164.6; IR (thin film) cm^{-1}: 3201, 2966, 1731, 1666, 1443, 1331, 1249. The submitters report an optical rotation for **3** of $[\alpha]_D$ +434° (EtOAc, c 1.70). Enantiomeric purity of **3** has been established by acylation of **3** with (S)-O-methylmandelic acid chloride followed by ^1H NMR analysis. Integration of the GC traces reveals that the resulting diastereomeric purity is equivalent to the enantiomeric purity of the mandelate (99%).

10. All glassware was oven dried at 120°C prior to use. Diethylamine was distilled from calcium hydride. Palladium(II) acetate, $Pd(OAc)_2$ (53 mg, 0.24), (cancer

suspect agent), without added triarylphosphine, gave **4** at a slightly lower yield as compared to Pd(PPh₃)₄.

11. During the course of the reaction, the solution is black. When the reaction is complete, the solution is clear brown with some black precipitate.

12. Some decomposition of the product, which appeared to be the major side reaction in this procedure, may occur in this step. The submitters found that decomposition is significant when the temperature is $\geq 20°C$, and the desired material is left in contact with the basic solution for extended periods of time. This procedure is the most efficient for larger amounts of material. For smaller amounts, base can be added to the aqueous layer in a separatory funnel and the resulting solution can be extracted with methylene chloride. In many cases the submitters have obtained material directly from this extraction that was suitable for further transformations.

13. The compound appears to be capable of isolation by sublimation, but this has not been checked. The submitters report obtaining **4** having mp 123-125°C and $[\alpha]_D$ -47° (CHCl₃, c 3.3). The reasons for the discrepancy in rotation have not been determined. The spectral properties of pure **4** are as follows: ¹H NMR (400 MHz, CDCl₃) δ: 1.27 (s, 9 H), 2.35 (dd, 1 H, J = 16.5, 12.0), 2.72 (dd, 1 H, J = 16.5, 5.7), 3.80 (s, 1 H); 4.71 (dd, 1 H, J = 12.0, 5.7), 6.89 (d, 2 H, J = 9), 7.3 (d, J = 9), 8.66 (s (br), 1 H); ¹³C NMR (100 MHz, CDCl₃) δ: 27.7, 37.6, 55.4, 56.7, 114.0, 127.5, 134.8, 158.8, 160.2, 171.7; IR (KBr) cm⁻¹: 3237, 3000, 1702, 1662, 1508 1254, 1135, 920, 832.

14. The properties of pure saturated heterocycle are as follows: mp 136-138°C, $[\alpha]_D$ +27.67° (CH₂Cl₂, c 1.2); ¹H NMR (400 MHz, CDCl₃) δ: 0.99 (s, 9 H), 2.33-2.72 (m, 2 H), 3.82 (s, 3 H), 3.97-4.04 (m, 1 H), 4.12 (s, 1 H), 6.03 (s (br), 1 H), 6.9 (d, 2 H, J = 10), 7.3 (d, 2 H, J = 10); ¹³C NMR (100 MHz, CDCl₃) δ: 24.9, 34.5, 39.8, 55.0, 55.4, 76.0, 114.1, 127.4, 134.2, 159.2, 171.6; IR (KBr) cm⁻¹: 3190, 2954, 1655, 1514, 1472, 1243, 1173.

15. Further purification including removal of the slight yellow color could be effected by recrystallization from boiling water. Physical properties of the purified amino acid product are as follows: mp 239°C (dec), $[\alpha]_D$ -4.35° (1 N HCl, c 1.86); ^{1}H NMR (400 MHz, D_2O) δ: 3.02-3.25 (m, 2 H), 3.83 (s, 3 H), 4.90 (m, 1 H), 7.03 (d, 2 H, J = 10), 7.45 (d, 2 H, J = 10); ^{13}C NMR (100 MHz, D_2O) δ: 38.7, 52.1, 56.6, 115.9, 128.7, 129.8, 160.8, 174.5; IR (KBr) cm^{-1}: 2960, 2364, 2150, 1615, 1518, 1404, 1250, 1184.

Waste Disposal Information

All toxic materials were disposed of in accordance with "Prudent Practices for Disposal of Chemicals from Laboratories"; National Academic Press; Washington, DC, 1983.

3. Discussion

Chemical methods for the production of enantiomerically pure α-amino acids have been extensively investigated in recent years.[3] Conversely, there are relatively few methods for the synthesis of chiral, nonracemic β-amino acids,[4] although there is considerable interest in these compounds as precursors to β-lactams,[5,6] as components of natural products,[7] and as reactive molecules in their own right.[8]

The procedure given here stems from our synthetic effort toward (+)-jasplakinolide[9-11] that contains the β-amino acid (R)-β-tyrosine.[12] This protocol permits the introduction of the desired *carbon substitutent* at the β-site in an enantioselective manner. This approach contrasts with previous methodologies, which develop the chiral center via conjugate addition of an amine to an α,β-unsaturated system,[13] reduction of a C=C or C=N functionality,[14] or C-C bond formation involving imines and enolate derivatives.[15]

The key element in this technology is heterocycle **3**. Although asparagine has been cyclized with acetone,[16] cyclocondensation with aldehydes has not been described in detail, except for tetrahydropyrimidinone formation using formaldehyde.[17] Electrochemical oxidative decarboxylation, on the other hand, is a well-known procedure[18] with broad applicability.[19] Our initial synthesis of **3** was accomplished with lead(IV) acetate;[20,21] the two-step method described here is cleaner and the cost and hazards of using and disposing of lead are precluded. Furthermore, the acid catalyst is easily recyclable. The two enantiomers of **3** are easily prepared from the corresponding enantiomers of asparagine which are readily available and inexpensive. The intermediate **3** and analogues are highly crystalline, stable, and possess large specific rotations which allows determination of the enantiomeric purity.

In addition, the protocol for the transformation of **3** to **4** has been modified from our original publication. We have found that the use of Et_2NH instead of triethylamine (Et_3N) as base gives a more reproducible reaction and, at the same time, all but eliminates the formation of biaryl material as a side product. The synthesis of **4** with Et_3N as base has been the subject of a mechanistic study and our results have been communicated.[21] Whether or not the mechanism is as we propose when Et_2NH is employed is not known.

Recently, we have demonstrated that **3** functions as an attractive chiral auxiliary for the synthesis of scalemic α-substituted carboxylic acids.[22] In addition, we have extended the scope of the β-amino acid synthesis to include alkyl groups through a protected derivative of heterocycle **3**.[23] Thus, a unified approach to the synthesis of enantiomerically pure β-amino acids has been developed from **3**. Recent efforts by Seebach, Juaristi, and co-workers[24] on a saturated analog of **3** have developed routes to products substituted at C_5 (α to the carbonyl).

1. Department of Chemistry and Biochemistry, University of California, Santa Cruz, CA 95064. This work was supported in part by the American Cancer Society and by the donors of the Petroleum Research Fund, administered by the American Chemical Society.

2. This procedure is taken from the work of Politzer, I. R.; Meyers, A. I. *Org. Synth., Coll. Vol. VI* **1988**, 905-909.

3. For recent developments in the area of α-amino acid synthesis, see O'Donnell, M. J., Ed. "Symposia-in-Print", *Tetrahedron* **1988**, *44*, 5253-5614.

4. (a) Drey, C. N. C. in "Chemistry and Biochemistry of the Amino Acids;" Barrett, G. C., Ed.; Chapman and Hall: New York, 1985; pp. 25-54; (b) Griffith, O. W. *Annu. Rev. Biochem.* **1986**, *55*, 855-878.

5. (a) Kobayashi, S.; Iimori, T.; Izawa, T.; Ohno, M. *J. Am. Chem. Soc.* **1981**, *103*, 2406-2408; (b) Shono, T.; Tsubata, K.; Okinaga, N. *J. Org. Chem.* **1984**, *49*, 1056-1059; (c) Liebeskind, L. S.; Welker, M. E.; Fengl, R. W. *J. Am. Chem. Soc.* **1986**, *108*, 6328-6343; (d) Davies, S. G.; Dordor-Hedgecock, I. M.; Sutton, K. H.; Walker, J. C. *Tetrahedron Lett.* **1986**, *27*, 3787-3790; (e) Kim, S.; Chang, S. B.; Lee, P. H. *Tetrahedron Lett.* **1987**, *28*, 2735-2736.

6. β-Lactams are also useful synthons. See (a) Ojima, I.; Chen, H.-J. C.; Nakahashi, K. *J. Am. Chem. Soc.* **1988**, *110*, 278-281; (b) Wasserman, H. H. *Aldrichim. Acta* **1987**, *20*, 63-74.

7. Some examples of natural products containing β-amino acids are (a) Blasticidin S: Yonehara, H.; Otake, N. *Tetrahedron Lett.* **1966**, 3785; (b) Bleomycin: Hecht, S. M. *Acc. Chem. Res.* **1986**, *19*, 383-391; (c) Bottromycin: Waisvisz, J. M.; van der Hoeven, M. G.; TeNijenhuis, B. *J. Am. Chem. Soc.* **1957**, *79*, 4524-4527; Nakamura, S.; Chikaike, T.; Yonehara, H.; Umezawa, H. *Chem. Pharm. Bull.* **1965**, *13*, 599-602; (d) Edeine: Hettinger, T. P.; Craig, L. C. *Biochemistry* **1968**, *7*, 4147-4153; (e) Scytonemin A: Helms, G. L.; Moore, R. E.; Niemczura,

W. P.; Patterson, G. M. L.; Tomer, K. B.; Gross, M. L. *J. Org. Chem.* **1988**, *53*, 1298-1307.

8. For example, see Wasserman, H. H.; Brunner, R. K.; Buynak, J. D.; Carter, C. G.; Oku, T.; Robinson, R. P. *J. Am. Chem. Soc.* **1985**, *107*, 519-521.

9. Isolation: (a) Crews, P.; Manes, L. V.; Boehler, M. *Tetrahedron Lett.* **1986**, *27*, 2797-2800; (b) Zabriskie, T. M.; Klocke, J. A.; Ireland, C. M.; Marcus, A. H.; Molinski, T. F.; Faulkner, D. J.; Xu, C.; Clardy, J. C. *J. Am. Chem. Soc.* **1986**, *108*, 3123-3124; (c) Braekman, J. C.; Daloze, D.; Moussiaux, B.; Riccio, R. *J. Nat. Prod.* **1987**, *50*, 994-995.

10. Chu, K. S.; Negrete, G. R.; Konopelski, J. P. *J. Org. Chem.* **1991**, *56*, 5196-5202.

11. Total synthesis: (a) Grieco, P. A.; Hon, Y. S.; Perez-Medrano, A. *J. Am. Chem. Soc.* **1988**, *110*, 1630-1631. Synthesis of fragments: (b) Schmidt, U.; Siegel, W.; Mundinger, K. *Tetrahedron Lett.* **1988**, *29*, 1269-1270; (c) Kato, S.; Hamada, Y.; Shioiri, T. *Tetrahedron Lett.* **1988**, *29*, 6465-6466. Conformation: (d) Inman, W.; Crews, P. *J. Am. Chem. Soc.* **1989**, *111*, 2822-2829.

12. Parry, R. J.; Kurylo-Borowska, Z. *J. Am. Chem. Soc.* **1980**, *102*, 836-837.

13. (a) Furukawa, M.; Okawara, T.; Terawaki, Y. *Chem. Pharm. Bull.* **1977**, *25*, 1319-1325; (b) d'Angelo, J.; Maddaluno, J. *J. Am. Chem. Soc.* **1986**, *108*, 8112-8114; (c) Baldwin, J. E.; Harwood, L. M.; Lombard, M. J. *Tetrahedron* **1984**, *40*, 4363-4370; (d) Feringa, B. L.; de Lange, B. *Tetrahedron Lett.* **1988**, *29*, 1303-1306; (e) Baldwin, S. W.; Aubé, J. *Tetrahedron Lett.* **1987**, *28*, 179-182; (f) Uyehara, T.; Asao, N.; Yamamoto, Y. *J. Chem. Soc., Chem. Commun.* **1989**, 753-754; (g) Matsunaga, H.; Sakamaki, T.; Nagaoka, H.; Yamada, Y. *Tetrahedron Lett.* **1983**, *24*, 3009-3012; (h) Davies, S. G.; Ichihara, O. *Tetrahedron: Asymmetry* **1991**, *2*, 183-186; (i) Hawkins, J. M.; Lewis, T. A. *J. Org. Chem.* **1992**, *57*, 2114-2121.

14. (a) Furukawa, M.; Okawara, T.; Noguchi, Y.; Terawaki, Y. *Chem. Pharm. Bull.* **1979**, *27*, 2223-2226; (b) Melillo, D. G.; Cvetovich, R. J.; Ryan, K. M.; Sletzinger, M. *J. Org. Chem.* **1986**, *51*, 1498-1504; (c) Chiba, T.; Ishizawa, T.; Sakaki, J.; Kaneko, C. *Chem. Pharm. Bull.* **1987**, *35*, 4672-4675; (d) Lubell, W. D.; Kitamura, M.; Noyori, R. *Tetrahedron: Asymmetry* **1991**, *2*, 543-544; (e) Potin, D.; Dumas, F.; d'Angelo, J. *J. Am. Chem. Soc.* **1990**, *112*, 3483-3486.

15. (a) Furakawa, M.; Okawara, T.; Noguchi, Y.; Terawaki, Y. *Chem. Pharm. Bull.* **1979**, *27*, 2795-2801; (b) Otsuka, M.; Yoshida, M.; Kobayashi, S.; Ohno, M.; Umezawa, Y.; Morishima, H. *Tetrahedron Lett.* **1981**, *22*, 2109-2112; (c) Yamasaki, N.; Murakami, M.; Mukaiyama, T. *Chem. Lett.* **1986**, 1013-1016; (d) Hatanaka, M. *Tetrahedron Lett.* **1987**, *28*, 83-86; (e) Shono, T.; Kise, N.; Sanda, F.; Ohi, S.; Tsubata, K. *Tetrahedron Lett.* **1988**, *29*, 231-234; (f) Kunz, H.; Schanzenbach, D. *Angew. Chem., Int. Ed. Engl.* **1989**, *28*, 1068-1069; (g) Gennari, C.; Venturini, I.; Gislon, G.; Schimperna, G. *Tetrahedron Lett.* **1987**, *28*, 227-230; (h) Davis, F. A.; Reddy, R. T.; Reddy, R. E. *J. Org. Chem.* **1992**, *57*, 6387-6389; (i) Laschat, S.; Kunz, H. *J. Org. Chem.* **1991**, *56*, 5883-5889; (j) Corey, E. J.; Decicco, C. P.; Newbold, R. C. *Tetrahedron Lett.* **1991**, *32*, 5287-5290; (k) Hua, D. H.; Miao, S. W.; Chen, J. S.; Iguchi, S. *J. Org. Chem.* **1991**, *56*, 4-6.

16. Hardy, P. M.; Samworth, D. J. *J. Chem. Soc., Perkin Trans. I* **1977**, 1954-1960.

17. French, D.; Edsall, J. T. *Adv. Prot. Chem.* **1945**, *2*, 277-355.

18. Seebach, D.; Charczuk, R.; Gerber, C.; Renaud, P.; Berner, H.; Schneider, H. *Helv. Chim. Acta* **1989**, *72*, 401-425.

19. Torii, S.; Tanaka, H. In "Organic Electrochemistry: An Introduction and Guide;" 3rd ed., Lund, H.; Baizer, M. M., Eds.; Marcel Deker: New York, 1991; pp. 535-579.

20. Konopelski, J. P.; Chu, K. S.; Negrete, G. R. *J. Org. Chem.* **1991**, *56*, 1355-1357.

21. Chu, K. S.; Negrete, G. R.; Konopelski, J. P.; Lakner, F. J.; Woo, N.-T.; Olmstead, M. M. *J. Am. Chem. Soc.* **1992**, *114*, 1800-1812.

22. Negrete, G. R.; Konopelski, J. P. *Tetrahedron: Asymmetry* **1991**, *2*, 105-108.

23. Chu, K. S.; Konopelski, J. P. *Tetrahedron* **1993**, *49*, 9183.

24. (a) Juaristi, E.; Quintana, D.; Lamatsch, B.; Seebach, D. *J. Org. Chem.* **1991**, *56*, 2553-2557; (b) Juaristi, E.; Escalante, J.; Lamatsch, B.; Seebach, D. *J. Org. Chem.* **1992**, *57*, 2396-2398; (c) Juaristi, E.; Quintana, D. *Tetrahedron: Asymmetry* **1992**, *3*, 723-726.

Appendix
Chemical Abstracts Nomenclature (Collective Index Number); (Registry Number)

2-tert-Butyl-1-carbomethoxy-2,3-dihydro-4(1H)-pyrimidinone: 1(2H)-Pyrimidine-carboxylic acid, 2-(1,1-dimethylethyl)-3,4-dihydro-4-oxo-, methyl ester, (S)- or (R)- (12); (S)-, (131791-75-8); (R)- (131791-81-6)

(R)-3-Amino-3-(p-methoxyphenyl)propionic acid: Benzenepropanoic acid, β-amino-4-methoxy-, (R)- (12); (131690-57-8)

L-Asparagine monohydrate: L-Asparagine (8,9); (70-47-3)

Pivalaldehyde (8); Propanal, 2,2-dimethyl- (9); (630-19-3)

Methyl chloroformate: Formic acid, chloro-, methyl ester (8); Carbonochloridic acid, methyl ester (9); (79-22-1)

Triethylamine (8); Ethanamine, N,N-diethyl- (9); (121-44-8)

(R)-2-tert-Butyl-6-(4-methoxyphenyl)-5,6-dihydro-4(1H)-pyrimidinone: 4(1H)-Pyrimidinone, 2-(1,1-dimethylethyl)-5,6-dihydro-6-(4-methoxyphenyl)-, (R)- (12); (131791-83-8; (S)- (131791-77-0)

4-Iodoanisole: Anisole, p-iodo- (8): Benzene, 1-iodo-4-methoxy- (9); (696-62-8)

Diethylamine (8); Ethanamine, N-ethyl- (9); (109-89-7)

Tetrakis(triphenylphosphine)palladium(0): Palladium, tetrakis(triphenylphosphine)- (8); Palladium, tetrakis(triphenylphosphine)-, (T-4)- (9); (14221-01-3)

Sodium borohydride: Borate (1-), tetrahydro-, sodium (8,9); (16940-66-2)

4-KETOUNDECANOIC ACID

(Undecanoic acid, 4-oxo-)

A.

$$\text{dihydrofuran} \xrightarrow[\text{2. } CH_3(CH_2)_6I]{\text{1. } t\text{-BuLi, } -78°C \rightarrow 0°C} \text{2-heptyldihydrofuran}$$

B.

$$\text{2-heptyldihydrofuran}—C_7H_{15} \xrightarrow[0°C]{H_2CrO_4} C_7H_{15}\text{C(O)CH}_2\text{CH}_2\text{CO}_2\text{H}$$

Submitted by M. A. Tschantz, L. E. Burgess, and A. I. Meyers.[1]

Checked by T. M. Kamenecka and L. E. Overman.

1. Procedure

A. *2-Heptyldihydrofuran.* An oven-dried, 1-L, round-bottomed flask, equipped with a magnetic stirring bar, and fitted with a rubber septum, is flushed with argon and charged with 8.00 mL of 2,3-dihydrofuran (105.8 mmol) (Note 1) and 600 mL of dry tetrahydrofuran (THF) (Note 2). The solution is stirred and cooled to -78°C using an external acetone-dry ice bath. After stirring at -78°C for 30 min, 50.0 mL of a solution of t-butyllithium in pentane (2.60 M in pentane, 130 mmol) (Note 3) is added dropwise via syringe over 60 min. The acetone-dry ice bath is replaced by an ice-water bath for 30 min, at which time the reaction mixture is recooled to -78°C. A solution of 17.35 mL of 1-iodoheptane (105.8 mmol) (Note 4) in 30 mL of dry tetrahydrofuran is added dropwise via syringe, and the resulting solution is allowed to warm to room temperature and stir for 60 min. The solution is recooled to 0°C and quenched by the careful addition of 100 mL of 50% aqueous ammonium chloride. The contents are transferred to a 1-L separatory funnel, and the organic phase separated. The aqueous phase is extracted three times with 100-mL portions of pentane-ether (1:1 v/v), and the combined organic phases are dried over Na_2SO_4, filtered, and concentrated under reduced pressure without heating to afford crude 2-heptyldihydrofuran (17.77 g, 100%) as a yellow oil (Note 5). This crude material is used directly in the next step.

B. *4-Ketoundecanoic acid.* A 2-L, round-bottomed flask, equipped with a magnetic stirring bar and a 250-mL addition funnel, is charged with 17.77 g of the crude 2-heptyldihydrofuran (105.8 mmol) and 400 mL of tetrahydrofuran (Note 6). The resulting solution is cooled to 0°C using an external ice bath, and 117.5 mL of a 2.7 M solution of aqueous chromic acid (317.4 mmol) (Note 7) is added dropwise *via* the addition funnel over 90 min, and the solution is allowed to stir overnight (Note 8). The reaction mixture is diluted with 300 mL of diethyl ether and 300 mL of water, and the resulting mixture is allowed to stir vigorously for 30 min. The reaction mixture is transferred to a separatory funnel, and the organic phase is separated (Note 9). The aqueous phase is extracted four times with 200-mL portions of diethyl ether. The combined organic phases are washed with three times with 100-mL portions of water (Note 10), followed by extraction three times with 150-mL portions of 10% aqueous sodium hydroxide solution. The combined basic extracts are acidified with 6 N hydrochloric acid to pH ~1 (*Caution:* Exothermic reaction!). The cloudy mixture is extracted 4 times with 150 mL of dichloromethane, and the combined organic extracts are dried over $MgSO_4$, followed by concentration under reduced pressure to afford crude 4-ketoundecanoic acid (12.4-14.7 g, 59-69%) as a white solid: mp 74-77°C (Note 11).

2. Notes

1. 2,3-Dihydrofuran was purchased from Fluka Chemical Corporation and used without further purification.

2. Tetrahydrofuran was freshly distilled over sodium/benzophenone prior to use.

3. t-Butyllithium was purchased from Lithco Chemical Corporation (the checkers used t-butyllithium purchased from Aldrich Chemical Company, Inc.) and titrated prior to use.

4. 1-Iodoheptane was purchased from Fluka Chemical Corporation and passed through activated basic aluminum oxide prior to use.

5. Because of the product's volatility, a hot water bath should not be used during solvent evaporation. GC and GC/MS analysis of an aliquot indicate that the product ranges in purity from 75-95% with unreacted 1-iodoheptane also present. The addition of 0.5 equiv of hexamethylphosphoramide (HMPA) prior to addition of the iodoheptane was found to improve the yield of this alkylation. The addition of 0.5 to 2.0 equiv of 1,3-dimethyl-3,4,5,6-tetrahydro-2(1H)-pyrimidinone (DMPU) did not improve the yield. The checkers found the following GC conditions useful for monitoring the alkylation reaction: initial column temperature, 40°C; heating increment, 10°C/min; iodoheptane R_f = 3.3 min, product R_f = 5.7 min. Column specifications were as follows: SPB-1 (stationary phase), fused silica gel capillary column, 30 m x 0.32 mm ID, 0.25-μm film thickness.

6. Acetone is commonly used as a solvent in Jones oxidations; however, the desired keto acid tends to be retained by the chromium salts during work-up. A benzene/THF solution has also been employed for the oxidation, but this modification did not seem to have much affect on the overall yield.

7. Aqueous chromic acid solution (the Jones reagent) was prepared according to "Reagents in Organic Syntheses", Fieser & Fieser; J. Wiley, 1967; Vol. 1, p. 142.

8. Chromium salts may precipitate upon addition of the Jones reagent. A minimal amount of water may be added to dissolve them.

9. Additional water may be added to facilitate separation of layers.

10. At this point, it is important to remove as much of the blue chromium impurities from the organic phase as possible.

11. The product may be recrystallized from hexanes to mp 79-80°C. The spectral data of the recrystallized material are as follows: ^{1}H NMR (500 MHz, CDCl$_3$) δ: 0.86 (t, 3 H, J = 7.0), 1.30 (s, 8 H), 1.57 (quintet, 2 H, J = 7.3), 2.43 (t, 2 H, J = 7.5), 2.61 (dd, 2 H, J = 6.23, 5.87), 2.71 (dd, 2 H, J = 6.60, 6.23); ^{13}C NMR (125 MHz, CDCl$_3$) δ: 14.0, 22.6, 23.8, 27.7, 29.0, 29.1, 31.6, 36.7, 42.7, 178.5, 208.9; IR (thin film) cm^{-1}: 3056, 2987, 2957, 2933, 2873, 2858, 1710, 1421, 1265, 748. Other electrophiles were used to give the corresponding keto acids as shown below:

Electrophile	Keto Acid Yield
C_4H_9I	64%
$C_{11}H_{23}I$	82%
$C_6H_5CH_2Br$	62%

Waste Disposal Information

All toxic materials were disposed of in accordance with "Prudent Practices for Disposal of Chemicals from Laboratories"; National Academic Press; Washington, DC, 1983.

3. Discussion

Because of a long standing research program centering on chiral bicyclic lactams, the submitter's laboratories have required a variety of 4-keto acids as precursors to these versatile intermediates.[2] Generally these acids are useful for the preparation of many important compounds.[3] A general route to 4-keto acids is the Larson procedure involving the C-silylation of butyrolactone followed by Grignard addition, elimination and in situ oxidation.[4] However, the major disadvantage to this process is the need for stoichiometric amounts of the expensive chlorodiphenylmethylsilane.

The procedure described here involves the metallation of dihydrofuran and subsequent alkylation with an alkyl iodide (bromides are much less reactive).[5] The resulting substituted dihydrofuran, the intermediate postulated in the Larson procedure, is then treated with chromic acid to hydrolyze the enol ether and oxidize the resulting primary alcohol to the corresponding carboxylic acid as in the Larson procedure. The isolated oxidation product is of suitable purity for subsequent reactions, but if necessary, recrystallization from hexanes is readily accomplished .

1. Department of Chemistry, Colorado State University, Fort Collins, CO 80523.

2. (a) Romo, D.; Meyers, A. I. *Tetrahedron* **1991**, *47*, 9503; (b) Meyers, A. I.; Berney, D. *Org. Synth.* **1990**, *69*, 55.

3. (a) Ellison, R. A. *Synthesis* **1973**, 397; (b) Ho, T.-L. *Synth. Commun.* **1974**, *4*, 265; (c) Stetter, H.; Schreckenberg, M. *Angew. Chem., Int. Ed. Engl.* **1973**, *12*, 81; (d) Nakamura, E.; Kuwajima, I. *J. Am. Chem. Soc.* **1984**, *106*, 3368; (e) Shimada, J.-i.; Hashimoto, K.; Kim, B. H.; Nakamura, E.; Kuwajima, I. *J. Am. Chem. Soc.* **1984**, *106*, 1759; (f) Tamaru, Y.; Ochiai, H.; Nakamura, T.; Tsubaki, K.; Yoshida, Z.-i. *Tetrahedron Lett.* **1985**, *26*, 5559; (g) Nakamura, E.;

Kuwajima, I. *Tetrahedron Lett.* **1986**, *27*, 83; (h) Sato, T.; Okazaki, H.; Otera, J.; Nozaki, H. *J. Am. Chem. Soc.* **1988**, *110*, 5209 and references therein; (i) Aoki, S.; Fujimura, T.; Nakamura, E.; Kuwajima, I. *Tetrahedron Lett.* **1989**, *30*, 6541.

4. (a) Larson, G. L.; Fuentes, L. M. *J. Am. Chem. Soc.* **1981**, *103*, 2418; (b) Fuentes, L. M.; Larson, G. L. *Tetrahedron Lett.* **1982**, *23*, 271; (c) Betancourt de Perez, R. M.; Fuentes, L. M.; Larson, G. L.; Barnes, C. L.; Heeg, M. J. *J. Org. Chem.* **1986**, *51*, 2039. For a related procedure, see (d) Larson, G. L.; Montes de Lopez-Cepero, I.; Mieles, L. R. *Org. Synth., Coll. Vol. VIII* **1993**, 474; (e) Viso, M.; Reid, J. R.; U.S. Patent 5 103 047, 1992; *Chem. Abstr.* **1992**, *116*, 235086k.

5. (a) Schlosser, M.; Schaub, B.; Spahic, B.; Sleiter, G. *Helv. Chim. Acta* **1973**, *56*, 2166; (b) Boeckman, Jr., R. K.; Bruza, K. J. *Tetrahedron* **1981**, *37*, 3997; (c) Kocienski, P.; Wadman, S.; Cooper, K. *J. Org. Chem.* **1989**, *54*, 1215.

Appendix
Chemical Abstracts Nomenclature (Collective Index Number); (Registry Number)

4-Ketoundecanoic acid: Undecanoic acid, 4-oxo- (8,9); (22847-06-9)

2,3-Dihydrofuran: Furan, 2,3-dihydro- (8,9); (1191-99-7)

tert-Butyllithium: Lithium, tert-butyl- (8); Lithium, (1,1-dimethylethyl)- (9); (594-19-4)

1-Iodoheptane: Heptane, 1-iodo- (8,9); (4282-40-0)

Sodium dichromate dihydrate: *Highly toxic.* CANCER SUSPECT AGENT: Chromic acid, disodium salt, dihydrate (8,9); (7789-12-0)

Sulfuric acid (8,9); (7664-93-9)

S-(-)-5-HEPTYL-2-PYRROLIDINONE.   CHIRAL BICYCLIC LACTAMS AS TEMPLATES FOR PYRROLIDINES AND PYRROLIDINONES

(2-Pyrrolidinone, 5-heptyl-, (S)-)

Submitted by M. A. Tschantz, L. E. Burgess, and A. I. Meyers.[1]

Checked by J. Madalengoitia and L. E. Overman.

1. Procedure

A. (+)-3-Phenyl-5-oxo-7a-heptyl-2,3,5,6,7,7a-hexahydropyrrolo[2,1-b]oxazole.
A clean, 500-mL, round-bottomed flask, equipped with a magnetic stirring bar and a
Dean-Stark trap, is charged with 12.5 g of 4-ketoundecanoic acid (62.4 mmol) (Note
1), 8.56 g of (S)-(+)-2-phenylglycinol (62.4 mmol) (Note 2), and 150 mL of toluene.
The solution is heated at reflux with stirring for 20 hr and then cooled to room
temperature. The solution is concentrated at reduced pressure, and the residual
yellow oil is purified by flash chromatography on silica gel (Note 3) employing 3:1
hexane/ethyl acetate as the eluent to provide 14.5-16.2 g of the bicyclic lactam (77-
86%) as a light yellow oil (Note 4).

B. (+)-N-[2-(1-Hydroxy-2-phenethyl)]-5-heptyl-2-pyrrolidinone. An oven-dried,
1-L, one-necked, round-bottomed flask, equipped with a magnetic stirring bar and
fitted with a rubber septum, is flushed with argon and charged with 14.1 g of the
bicyclic lactam (46.7 mmol) prepared above and 270 mL of dry dichloromethane (Note
5). The resulting solution is cooled to -78°C, using an external dry ice-acetone bath,
and stirred for 15 min. Then 13.4 mL of triethylsilane (84.1 mmol) (Note 6) and 93 mL
(93 mmol) of a 1.0 M solution of titanium tetrachloride in CH_2Cl_2 (Note 7) are
successively added dropwise via syringe under an argon atmosphere (Note 8). The
solution is stirred at -78°C for 1.5 hr and then allowed to warm to room temperature
and stir at that temperature for 1 hr. The mixture is recooled to 0°C and carefully
quenched with saturated ammonium chloride via syringe (*Caution: vigorous
reaction!*). The resulting solution is diluted with 120 mL of water, transferred to a 1-L
separatory funnel, and extracted with three 75-mL portions of dichloromethane. The
combined organic extracts are dried over sodium sulfate (Na_2SO_4), filtered, and
concentrated under reduced pressure to afford a crude mixture containing the product.
Purification of the product by flash chromatography, using 1:1 (v/v) hexane/ethyl

acetate as eluent, provides the hydroxypyrrolidinone (12.6 g, 89%) as a pale yellow oil (Note 9).

C.   S-(-)-5-Heptyl-2-pyrrolidinone.   An oven-dried, 1-L, two-necked, round-bottomed flask, equipped with a glass-covered stirring bar, is charged with 12.4 g of (+)-N-[2-(1-hydroxy-2-phenethyl)]-5-heptyl-2-pyrrolidinone (40.8 mmol), 24 mL of anhydrous ethyl alcohol (410 mmol), and 100 mL of dry tetrahydrofuran (THF). One neck is fitted with an acetone-dry ice cooled cold finger condenser bearing an argon inlet/outlet vented through a mineral oil bubbler, and the other neck with a second gas inlet valve. With stirring, the solution is cooled to -78°C using an external acetone-dry ice bath, and ammonia (~200 mL) is condensed into the flask through the gas inlet valve. The gas inlet is replaced with a stopper, the acetone-dry ice bath is removed, and 2.83 g of lithium wire (408 mmol) (Note 10) is added in small portions by removal of the stopper. The resulting blue solution is allowed to stir and reflux for 30 min (Note 11). The reaction mixture is carefully quenched with a small amount of ammonium chloride (solid) until disappearance of the blue color. The cold finger is removed, and the resulting solution is allowed to warm to room temperature over 4 hr (*Caution*: *ammonia is evolved*). The residual milky solution is concentrated under reduced pressure, and the remaining semisolid is diluted with water (100 mL) and dichloromethane (150 mL) and stirred for 30 min. The two layers are separated, and the aqueous layer is extracted four times with 50-mL portions of dichloromethane. The combined organic phases are dried over magnesium sulfate, filtered, and concentrated under reduced pressure. The yellow residue is subjected to flash chromatography (Notes 3 and 12), employing a 4:1 hexane/ethyl acetate solution as eluent, to afford 4.64 g (62%) of the desired pyrrolidinone as a pale yellow solid, mp 43-44°C (Note 13). *Caution: The minor reaction by-product, phenethyl alcohol, which has a distinct benzene-like odor, is an inhibitor of DNA and RNA synthesis in vitro.*

2. Notes

1. Tschantz, M. A.; Burgess, L. E.; Meyers, A. I. *Org. Synth.* **1995**, *73*, 215.

2. (S)-(+)-2-Phenylglycinol is commercially available, or may be prepared by reducing (S)-(+)-2-phenylglycine as follows: A 1-L, three-necked, round-bottomed flask is charged with 200 mL of dry THF under an argon atmosphere. Portionwise, 10.01 g of sodium borohydride (264.6 mmol) is added, followed by the dropwise addition of 65.1 mL of boron trifluoride etherate (528.7 mmol). The colorless suspension is stirred for 15 min, followed by portionwise addition of 20.00 g of (S)-(+)-2-phenylglycine (132.2 mmol). (*Caution: exotherm and gas evolution!*). The resulting suspension is heated at reflux for 12 hr and then is allowed to cool to room temperature followed by quenching with methanol until gas evolution ceases. The reaction mixture is concentrated under reduced pressure to yield a colorless solid that is taken up in 400 mL of 20% aqueous sodium hydroxide solution. The basic solution is extracted three times with 200-mL portions of dichloromethane, the combined organic layers are dried over sodium sulfate, and concentrated under reduced pressure to yield 14.72 g (81%) of (S)-(+)-2-phenylglycinol as a colorless solid: mp 72-74°C, $[\alpha]_D$ +32.2°; ^{1}H NMR (300 MHz, CDCl$_3$) δ: 3.52 (dd, 1 H, J = 10.7, 8.3), 3.71 (dd, 1 H, J = 10.7, 4.4), 4.01 (dd, 1 H, J = 8.2, 4.4), 7.23-7.35 (m, 5 H).

3. Merck 951 grade silica gel was used. Chromatography was performed in the manner described by Still[2] using ~20:1 (w/w) of silica gel to crude product.

4. The spectral data of the purified bicyclic lactam are as follows: ^{1}H NMR (300 MHz, CDCl$_3$) δ: 0.85 (t, 3 H, J = 6.4), 1.13-1.72 (m, 12 H), 2.13 (m, 1 H), 2.33 (ddd, 1 H, J = 13.4, 9.7, 2.5), 2.57 (ddd, 1 H, J = 17.3, 10.2, 2.5), 2.81 (m, 1 H), 4.05 (dd, 1 H, J = 8.7, 7.3), 4.62 (t, 1 H, J = 8.5), 5.17 (t, 1 H, J = 7.7), 7.20-7.36 (m, 5 H); ^{13}C NMR (75.5 MHz, CDCl$_3$) δ: 14.0, 22.5, 23.9, 29.1, 29.5, 30.9, 31.6, 33.3, 36.3, 57.5, 72.8, 102.7,

125.4, 127.3, 128.6, 140.1, 179.3; IR (thin film) cm^{-1}: 2954-2856, 2362, 1715, 1458, 1364, 1031, 699.

5. Dichloromethane was freshly distilled over calcium hydride.

6. Triethylsilane was purchased from Aldrich Chemical Company, Inc. and used without further purification.

7. Titanium tetrachloride was purchased from Fluka Chemical Corporation and used without further purification.

8. The checkers used an internal thermometer and maintained the reaction temperature between -78° and -70 °C during the addition.

9. The spectral data of the purified hydroxyalkyl lactam are as follows: ^{1}H NMR (300 MHz, CDCl$_3$, 5 mg/mL) δ: 0.86 (t, 3 H, J = 6.6), 1.05-1.40 (m, 11 H), 1.58 (m, 1 H), 1.75 (m, 1 H), 2.07 (m, 1 H), 2.40-2.59 (m, 2 H), 2.95 (s (br), 1 H), 3.33 (m, 1 H), 3.95 (dd, 1 H, J = 12.3, 3.3), 4.22 (dd, 1 H, J = 12.3, 7.8), 4.41 (dd, 1 H, J = 7.8, 3.3), 7.20-7.37 (m, 5 H); ^{13}C NMR (75.5 MHz, CDCl$_3$) δ: 14.0, 22.6, 24.1, 24.5, 29.1, 29.4, 31.3, 31.7, 32.3, 59.3, 62.3, 64.1, 127.2, 127.8, 128.7, 137.5, 176.8; IR (thin film) cm^{-1}: 3377 (br), 2959-2856, 1668, 1451, 1420, 1285, 1064, 702. The ^{1}H NMR signal for the proton of the OH group (δ 2.95) was found to be concentration dependent.

10. Lithium wire was purchased from Aldrich Chemical Company, Inc., and was cut into small pieces.

11. The solution should remain blue for the duration (30 min). More lithium may be added if necessary.

12. The final product is difficult to visualize during TLC analysis. Ninhydrin (5% w/v in ethanol) or ceric ammonium nitrate (5 g in 100 mL of aqueous 1 M H$_2$SO$_4$) are effective.

13. The spectral data for the purified pyrrolidinone are as follows: ^{1}H NMR (300 MHz, CDCl$_3$, 5 mg/mL) δ: 0.86 (t, 3 H, J = 6.4), 1.25-1.75 (m, 13 H), 2.20-2.35 (m, 3 H), 3.61 (quintet, 1 H, J = 6.9), 6.09 (s (br), 1 H); ^{13}C NMR (75.5 MHz, CDCl$_3$) δ: 14.0,

225

22.6, 25.8, 27.3, 29.1, 29.4, 30.2, 31.7, 36.7, 54.7, 178.2; IR (thin film) cm^{-1}: 3204 (br), 2956-2856, 1700, 1463, 1387, 1311, 1266. [α]$_D$ -9.7° (CHCl$_3$, c 1.1).

Waste Disposal Information

All toxic materials were disposed of in accordance with "Prudent Practices for Disposal of Chemicals from Laboratories"; National Academic Press; Washington, DC, 1983.

3. Discussion

Chiral bicyclic lactams are excellent precursors to a wide variety of chiral, non-racemic carbocycles including cyclopentenones, cyclohexenones, cyclopropanes, indanones, naphthalenones, and asymmetric keto acids.[3] Recently they have been applied to the synthesis of chiral, non-racemic pyrrolidines and pyrrolidinones,[4] that are medicinally and synthetically important molecules.[5] The three-step procedure described here provides an efficient route (overall yield: 46%) to (S)-5-heptyl-2-pyrrolidinone of high enantiomeric purity. The scheme below illustrates this reaction.

R	Yield of 1	Yield of 2	Yield of 3
cyclopentyl	83%	88%	89%
isobutyl	89%	85%	65%
isopropyl	84%	94%	81%

The first step is the formation of the chiral bicyclic lactam **1** involving the cyclodehydration of a 4-keto acid (see preceding procedure) and (S)-2-phenylglycinol. Interestingly only a single diastereomer (NMR) of **1** is formed. The second step, the crucial reduction of the N,O-ketal, probably proceeds via an N-acyl iminium ion in which allylic 1,3-interactions dictate the facial selectivity of hydride delivery.[4b,c] The resulting N-substituted 5-alkylpyrrolidinones **2** were typically obtained in >94% diastereomeric excess as determined by the isolation of each diastereomer via flash chromatography. The final step, cleavage of the chiral auxiliary, takes advantage of the benzylic C-N bond to expose the free amide by dissolving metal reduction. Although the chiral "auxiliary" is destroyed, the relatively low cost of both antipodes of phenylglycine make this procedure attractive as a preparative method for both antipodes of the pyrrolidinones.

1. Department of Chemistry, Colorado State University, Fort Collins, CO 80523.

2. Still, W. C.; Kahn, M.; Mitra, A. *J. Org. Chem.* **1978**, *43*, 2923.

3. (a) Romo, D.; Meyers, A. I. *Tetrahedron* **1991**, *47*, 9503 and references therein; (b) Meyers, A. I.; Berney, D. *Org. Synth.* **1990**, *69*, 55.

4. (a) Meyers, A. I.; Burgess, L. E. *J. Org. Chem.* **1991**, *56*, 2294; (b) Burgess, L. E.; Meyers, A. I. *J. Am. Chem.* **1991**, *113*, 9858; (c) Burgess, L. E.; Meyers, A. I. *J. Org. Chem.* **1992**, *57*, 1656 and references therein; (d) Meyers, A. I.; Snyder, L. *J. Org. Chem.* **1992**, *57*, 3814.

5.   In natural products: (a) Daly, J. W.; Spande, T. F. in "Alkaloids: Chemical and Biological Perspectives"; Pelletier, S. W.; Ed.; Wiley: New York, 1986; Vol. 4, Chapter 1; (b) Hart, D. J. in "Alkaloids: Chemical and Biological Perspectives"; Pelletier, S. W.; Ed.; Wiley: New York, 1988; Vol 6, Chapter 3; (c) Elbein, A. D.; Molyneux, R. J. "Alkaloids: Chemical and Biological Perspectives"; Pelletier, S. W.; Ed.; Wiley: New York, 1987, Vol. 5, Chapter 1; (d) Gellert, E. "Alkaloids: Chemical and Biological Perspectives"; Pelletier, S. W.; Ed.; Wiley: New York, 1987, Vol. 5, Chapter 2; (e) Hiemstra, H.; Speckamp, W. N. in "The Alkaloids"; Brossi, A., Ed.; Academic Press: New York, 1988, Vol. 32, Chapter 4; (f) Massiot, G.; Delaude, C. in "The Alkaloids"; Brossi, A., Ed.; Academic Press: New York, 1986; Vol 27, Chapter 3; (g) Jones, T. H.; Blum, M. S.; Fales, H. M. *Tetrahedron* **1982**, *38*, 1949; (h) Jones, T. H.; Blum, M. S.; Howard, R. W.; McDaniel, C. A.; Fales, H. M.; Dubois, M. B.; Torres, J. *J. Chemical Ecology* **1982**, *8*, 285; (i) Rüeger, H.; Benn, M. *Heterocycles* **1983**, *20*, 1331; (j) Stevens, R. V. *Accts. Chem. Res.* **1977**, *10*, 193; (k) Stevens, R. V. In "The Total Synthesis of Natural Products:, ApSimon, J., Ed.; Wiley: New York, 1977; Vol. 3, p. 439.   As medicinally important agents: (l) Rigo, B.; Fasseur, D.; Cherepy, N.; Couturier, D. *Tetrahedron Lett.* **1989**, *30*, 7057; (m) Nilsson, B. M.; Ringdahl, B.; Hacksell, U. *J. Med. Chem.* **1990**, *33*, 580; (n) Lundkvist, J. R. M.; Wistrand, L.-G.; Hacksell, U. *Tetrahedron Lett.* **1990**, *31*, 719; (o) Garvey, D. S.; May, P. D.; Nadzan, A. M. *J. Org. Chem.* **1990**, *55*, 936; (p) Heffner, R. J.; Joullié, M. M. *Tetrahedron Lett.* **1989**, *30*, 7021; (q) Silverman, R. B.; Nanavati, S. M. *J. Med. Chem.* **1990**, *33*, 931; (r) Bick, I. R. C.; Hai, M. A. in "The Alkaloids"; Brossi, A., Ed.; Academic Press: New York, 1985; Vol. 24, Chapter 3; (s) Aronstam, R. S.; Daly, J. W.; Spande, T. F.; Narayanan, T. K.; Albequerque, E. X. *Neurochemical Res.* **1986**, *11*, 1227; (t) Hino, M.; Nakayama, O.; Tsurumi, Y.; Adachi, K.; Shibata, T.; Terano, H.; Kohsaka, M.; Aoki, H.; Imanaka, H. *J. Antibiot.* **1985**, *38*,

926; (u) Humphries, M. J.; Matsumoto, K.; White, S. L.; Molyneux, R. J.; Olden, K. *Cancer Res.* **1988**, *48*, 1410; (v) Jimenez, A.; Vazques, D. in "Antibiotics", Hahn, F. E., Ed.; Springer Verlag: Berlin, 1979; Vol. 5, Part 2, pp. 1-19 and references cited therein. As chiral auxiliaries: (w) Whitesell, J. K. *Chem. Rev.* **1989**, *89*, 1581-1590; (x) See "Asymmetric Synthesis"; Morrison, J. D.; Ed.; Academic Press: New York, 1983, Vol. 2. As chiral bases: (y) Tomioka, K. *Synthesis* **1990**, 541 and references therein; (z) Cain, C. M.; Cousins, R. P. C.; Coumbarides, G.; Simpkins, N. S. *Tetrahedron* **1990**, *46*, 523; (aa) Cox, P. J.; Simpkins, N, S. *Tetrahedron: Asymmetry* **1991**, *2*, 1. As chiral ligands: (bb) Kagan, H. B. in "Asymmetric Synthesis"; Morrison, J. D , Ed.; Academic Press: New York, 1985; Vol. 5, Chapter 1; (cc) Doyle, M. P.; Pieters, R. J.; Martin, S. F.; Austin, R. E.; Oalmann, C. J.; Müller, P. *J. Am. Chem. Soc.* **1991**, *113*, 1423 and references cited therein.

Appendix

Chemical Abstracts Nomenclature (Collective Index Number); (Registry Number)

(S)-(-)-5-Heptyl-2-pyrrolidinone: 2-Pyrrolidinone, 5-heptyl-, (S)- (13); (152614-98-7)

(+)-3-Phenyl-5-oxo-7a-heptyl-2,3,5,6,7,7a-hexahydropyrrolo[2,1-b]oxazole: Pyrrolo-[2,1-b]oxazol-5(6H)-one, tetrahydro-7a-heptyl-3-phenyl-, (3R-cis)- (12); (132959-41-2)

4-Ketoundecanoic acid: Undecanoic acid, 4-oxo- (8,9); (22847-06-9)

(S)-(+)-2-Phenylglycinol: Phenethyl alcohol, β-amino-, L- (8,9); (20989-17-7)

Toluene (8); Benzene, methyl- (9); (108-88-3)

(+)-N-[2(1-Hydroxy-2-phenethyl)]-5-heptyl-2-pyrrolidinone: 2-Pyrrolidinone, 5-heptyl-1-(2-hydroxy-1-phenylethyl)-, [R-(R*,R*)]- (13); (139564-36-6)

Dichloromethane: Methane, dichloro- (8,9); (75-09-2)

Triethylsilane: Silane, triethyl- (8,9); (617-86-7)

Titanium tetrachloride: HIGHLY TOXIC: Titanium chloride (8,9); (7550-45-0)

Lithium (8,9); (7439-93-2)

(S)-(+)-2-Phenylglycine: Glycine, 2-phenyl-, L- (8); Benzeneacetic acid, α-amino-, (S)- (9); (2935-35-5)

Sodium borohydride: Borate (1-), tetrahydro-, sodium (8,9); (16940-66-2)

Boron trifluoride etherate: Ethyl ether, compd. with boron fluoride (1:1) (8); Ethane, 1,1'-oxybis-, compd. with trifluoroborane (1:1)

AN IMPROVED PREPARATION OF 3-BROMO-2(H)-PYRAN-2-ONE: AN AMBIPHILIC DIENE FOR DIELS-ALDER CYCLOADDITIONS

(2H-Pyran-2-one, 3-bromo-)

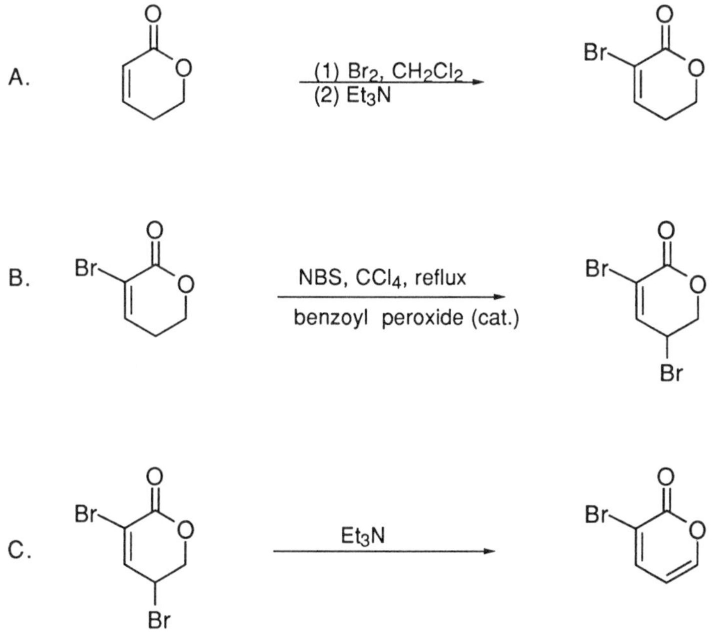

A.

(1) Br₂, CH₂Cl₂
(2) Et₃N

B.

NBS, CCl₄, reflux

benzoyl peroxide (cat.)

C.

Et₃N

Submitted by G. H. Posner, K. Afarinkia, and H. Dai.[1]
Checked by Alyx-Caroline Guevel and David J. Hart.

1. Procedure

A.   3-Bromo-5,6-dihydro-2(H)-pyran-2-one.   A 1-L, three-necked, round-bottomed flask, equipped with a magnetic stirring bar, a pressure-equalizing addition funnel, a drying tube that contains sodium hydroxide pellets, and a thermometer, is charged with 10.15 g (0.103 mol) of 5,6-dihydro-2(H)-pyran-2-one (Note 1) and 350

231

mL of methylene chloride (Note 2). The addition funnel is charged with a solution of 16.7 g (0.105 mol) of bromine (Note 3) in 130 mL of methylene chloride. The bromine solution is added over a period of 4 hr (Note 4) in 10-15-mL portions *(Caution! Exothermic reaction. External cooling may be necessary).* After the addition of the last portion of bromine, the reaction is analyzed by TLC and NMR (Notes 5 and 6). The resulting pale orange solution is stirred for 2 hr until the color has almost faded. The addition funnel is replaced by a rubber septum, and the reaction mixture is cooled by means of an ice bath. Via syringe, 15.0 mL (0.107 mol) of triethylamine is added through the rubber septum over 2 min *(Caution! Exothermic reaction).* The colorless solution is stirred for 40 min and then the contents of the flask are transferred to a 1-L separatory funnel and washed twice with 150 mL of water. The organic phase is dried over anhydrous sodium sulfate and filtered through a pad of silica gel (Note 7) and the pad rinsed with 700 mL of methylene chloride. The combined filtrates are concentrated at reduced pressure (20 mm) and the last traces of solvent are removed under high vacuum at ambient temperature to afford 16.3 g (89%) of 3-bromo-5,6-dihydro-2(H)-pyran-2-one as an amber-colored mobile liquid that solidifies when stored overnight at -4°C. This material is sufficiently pure for use in the next step. An analytically pure sample may be prepared by Kugelröhr distillation at 1.0 mm (pot temperature 100°C) (Note 8).

   B. 3,5-Dibromo-5,6-dihydro-2(H)-pyran-2-one. A flame-dried, 1-L, round-bottomed flask, equipped with a magnetic stirring bar and an efficient water-cooled condenser fitted with a nitrogen inlet, is placed under a nitrogen atmosphere and charged with 15.8 g (0.089 mol) of 3-bromo-5,6-dihydro-2(H)-pyran-2-one, 16.5 g (0.093 mol) of N-bromosuccinimide (Note 1), 0.8 g (3.3 mmol) of benzoyl peroxide (Note 1), and 455 mL of freshly distilled carbon tetrachloride. The reaction flask is immersed in a preheated oil bath (100°C) and the contents are refluxed vigorously for 4.5 hr (Notes 9 and 10); the reaction mixture is cooled and allowed to stand at room temperature for 4.5 hr. The precipitate (9.5 g) is collected by suction filtration, the filtrate is concentrated under reduced pressure at 50-60°C (20 mm), and the last traces of solvent are removed under high vacuum (1 mm) at room temperature over 2.5 hr affording 21.5 g (94%) of crude 3,5-dibromo-5,6-dihydro-2(H)-pyran-2-one as an amber liquid which is sufficiently pure for use in the next step (Note 11). This crude material may be further purified by chromatography (Note 12).

C. 3-Bromo-2(H)-pyran-2-one. A 1-L, one-necked, round-bottomed flask, equipped with a magnetic stirring bar, and rubber septum through which a needle-tipped inert gas line is inserted (vented through a mineral oil bubbler), is charged with 20.3 g of crude 3,5-dibromo-5,6-dihydro-2(H)-pyran-2-one, as prepared above, and 360 mL of methylene chloride (Note 1). Via syringe, 14.0 mL (0.1 mol) of triethylamine (Note 2) is added over a 5-min period at room temperature. During the addition, the reaction mixture becomes dark brown. After 24 hr, analysis by TLC and NMR indicates the reaction to be complete. The reaction mixture is transferred to a 1-L separatory funnel and washed three times with 140 mL of water. The organic phase is dried over anhydrous sodium sulfate, filtered and concentrated under reduced pressure at 50-60°C (20 mm). The last traces of solvent and triethylamine are removed by brief exposure (10 min) to high vacuum at ambient temperature (Note 13). The residue is purified by chromatography on a 5.0 cm-in-diameter column of 150 g of silica gel packed in hexane. The crude material is dissolved in a minimum amount of methylene chloride and applied to the column. Elution with 800 mL of 10% ethyl acetate in hexane followed by 1300 mL of 20% ethyl acetate in hexane, using analytical TLC to monitor fractions, affords in sequence 1.97 g (14%) of 5-bromo-2-pyrone as a light-brown solid with mp 54-56°C (lit.[4] mp 60-61°C) (Note 14), and 6.06 g (43%) of 3-bromo-2-pyrone as a tan solid with mp 59.5-61°C (lit.[3,5] mp 63.5-65°C), 36% yield overall from 5,6-dihydro-2(H)-pyran-2-one. (Notes 15 and 16).

2. Notes

1. All commercially available reagents were used as received without further purification. Benzoyl peroxide was 95% pure by iodometric titration.

2. All common reagents were dried according to recommended procedures[2] and were redistilled prior to use.

3. The weight of bromine should be measured by addition to a cooled, pre-weighed Erlenmeyer flask containing methylene chloride. *Caution: bromine is toxic and should always be handled in a well-ventilated fumehood.*

4. The rate of addition is determined by the temperature of the reaction mixture and the quality of the starting pyrone. In a duplicate run, starting from redistilled starting material, addition took only 2 hr.

5. TLC analysis was carried out on precoated plastic silica gel plates (with fluorescent indicator) using diethyl ether as eluant. With this eluant, the R_f of starting material is 0.50 and the R_f of 3,4-dibromo-3,4,5,6-tetrahydro-2(H)-pyran-2-one is 0.83. In addition to these two spots, a third spot, $R_f = 0.55$, corresponding to 3-bromo-5,6-dihydro-2(H)-pyran-2-one was detected. TLC analysis is not always reliable and it is better to analyze a small aliquot by NMR.

6. More bromine may be necessary. *Avoid addition of more bromine if the starting material has already been consumed.*

7. A pad of 100 g of flash silica gel (1.5 cm thick and 11 cm diameter) was used.

8. Distillation of 1.71 g of crude material afforded 1.50 g of pure 3-bromo-5,6-dihydro-2(H)-pyran-2-one as a colorless oil which solidifies at -4°C (mp 27-30°C), and which slowly turns yellow upon standing at room temperature. The spectral data for 3-bromo-5,6-dihydro-2(H)-pyran-2-one is as follows: ^{1}H NMR (400 MHz, $CDCl_3$) δ: 2.58 (dt, 2 H, J = 6.1, 4.6), 4.49 (t, 2 H, J = 6.1), 7.30 (t, 1 H, J = 4.6); ^{13}C NMR (62.5 MHz,

CDCl$_3$) δ: 26.5, 66.7, 113.3, 146.2, 159.1; IR (neat) cm^{-1}: 1733, 1615; Mass spectrum (m/z) (EI, 70 eV) 177.9 and 176.9 (M$^+$, 42). Anal. Calcd for C$_5$H$_5$O$_2$Br: C, 33.91; H, 2.85; Br, 45.16. Found C, 33.99; H, 2.88; Br, 45.08.

9. Shortly after refluxing begins, the reaction mixture turns dark orange; however, the color dissipates by the end of the reaction period.

10. Completion of the reaction was checked by TLC analysis (Note 5). The R$_f$ of the product is 0.60 with diethyl ether as eluant.

11. The crude material is contaminated with starting bromide, small amounts of succinimide and aromatic material, as well as other minor impurities as indicated by ^{1}H NMR.

12. Chromatography of 1.17 g of this material on 50 g of silica gel with elution by a gradient of 20-50% ethyl acetate in hexane gave 0.90 g (77%) of ~94% pure 3,5-dibromo-5,6-dihydro-2(H)-pyran-2-one as a yellow oil that crystallized on standing (mp 60.5-62.5°C; lit.[3] mp 64-65°C). The spectral data for 3,5-dibromo-5,6-dihydro-2(H)-pyran-2-one is as follows: ^{1}H NMR (400 MHz, CDCl$_3$) δ: 4.62-4.67 (m, 1 H), 4.76-4.81 (m, 2 H), 7.37 (d, 1 H, J = 5.7). This material was contaminated with 6% of 3-bromo-5,6-dihydro-2(H)-pyrone as indicated by ^{1}H NMR.

13. 3-Bromo-2-pyrone sublimes in high vacuum at relatively low temperature. Thus, the material should not be subjected to high vacuum for long periods of time.

14. The spectral data for 5-bromo-2-pyrone is as follows: ^{1}H NMR (400 MHz, CDCl$_3$) δ: 6.31 (dd, 1 H, J = 9.8, 1.1), 7.38 (dd, 1 H, J = 9.8, 2.7), 7.62 (dd, 1 H, J = 2.7, 1.1); ^{13}C NMR (50 MHz, CDCl$_3$) δ: 100.7, 117.4, 145.9, 149.6, 159.3.

15. The spectral data for 3-bromo-2-pyrone is as follows: ^{1}H NMR (400 MHz, CDCl$_3$) δ: 6.15 (dd, 1 H, J = 6.9, 5.0), 7.51 (dd, 1 H, J = 5.0, 1.9), 7.69 (dd, J = 6.9, 1.9); ^{13}C NMR (50 MHz, CDCl$_3$) δ: 106.5, 112.6, 144.0, 150.1, 158.0.

16. The checkers also isolated 0.7 g (3%) of 3,5-dibromo-2-pyrone as a brown solid, mp 55.5-58.5°C (lit.[4] mp 66-67°C), from the fractions preceding the 5-bromo-2(H)-pyrone.

Waste Disposal Information

All toxic materials were disposed of in accordance with "Prudent Practices for Disposal of Chemicals from Laboratories"; National Academic Press; Washington, DC, 1983.

3. Discussion

The procedure described above allows a more efficient and convenient synthesis of 3-bromo-2-pyrone than previously described in the literature.[5] This procedure avoids making and handling 2-pyrone, a sensitive compound.[6] The major by-product of the reaction is 5-bromo-2-pyrone. We postulate that the formation of this by-product results from prototropic migration in basic medium followed by elimination of HBr (Scheme 1).

Scheme 1

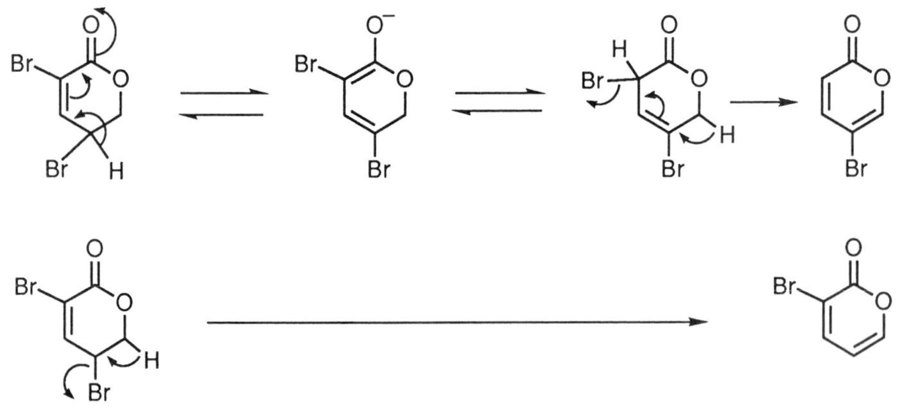

3-Bromo-2-pyrone is not only a valuable precursor for the synthesis of various 3-substituted 2-pyrones,[7] but it is also a reactive unsymmetrical diene.[8] 3-Bromo-2-pyrone undergoes Diels-Alder cycloadditions with a regioselectivity and stereoselectivity that is superior to that of 2-pyrone. Furthermore, 3-bromo-2-pyrone is a chameleon (i.e., ambiphilic) dienophile, undergoing cycloaddition to both electron deficient and electron rich dienophiles. The cycloadducts of bromopyrone with dienophiles are isolable and are useful in the synthesis of diastereomerically pure cyclohexene carboxylates (Scheme 2).[8]

Scheme 2

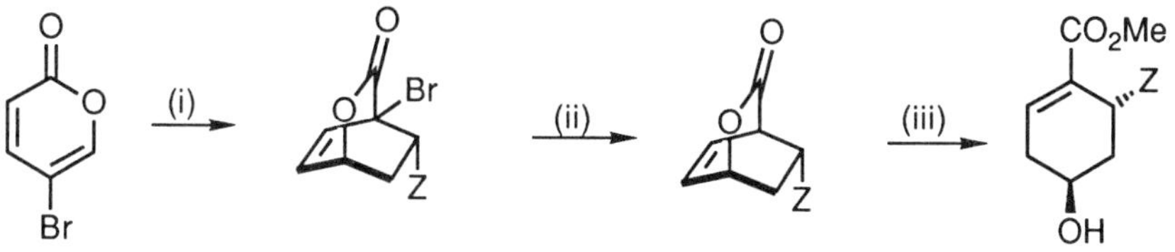

Z = CH$_2$OSiMe$_2$-t-Bu: (i) Acrolein (10 equiv), 90°C, 96 hr, 70%; then NaBH$_4$, MeOH, 0°C, 15 min (80%); then TSDMSTf, Et$_3$N, CH$_2$Cl$_2$, 30 min (90%); (ii) Bu$_3$SnH, AIBN, benzene, reflux 2 hr (99%); (iii) MeONa, MeOH, 0°C, 92%.

Z = OSiPh$_2$Me: (i) CH$_2$=CHOSiPh$_2$Me (10 equiv), 90°C, 140 hr, 44%; (ii) Bu$_3$SnH, AIBN, benzene, reflux 2 hr (47%); (iii) MeOLi, MeOH, 0°C, 90%.

1. Department of Chemistry, The Johns Hopkins University, Baltimore, MD 21218. We thank the NIH (GM-30052) for financial support.

2. Perrin, D. D.; Armarego, W. L. F.; Perrin, D. R. "Purification of Laboratory Chemicals," 2nd ed.; Pergamon Press: New York, 1980.

3. Bock, K.; Pedersen, C.; Rasmussen, P. *Acta Chem. Scan.* **1975**, *29B*, 389.

4. Pirkle, W. H.; Dines, M. *J. Org. Chem.* **1969**, *34*, 2239.

5. Pirkle, W. H.; Dines, M. *J. Heterocyclic Chem.* **1969**, *6*, 1.

6.   Nakagawa, M.; Saegusa, J.; Tonozuka, M.; Obi, M.; Kiuchi, M.; Hino, T.; Ban, Y. *Org. Synth., Coll. Vol. VI* **1988,** 462.

7.   (a) Posner, G. H.; Harrison, W.; Wettlaufer, D. G. *J. Org. Chem.* **1985,** *50,* 5041; (b) Posner, G. H.; Nelson, T. D.; Kinter, C. K.; Johnson, N. *J. Org. Chem.* **1992,** *57,* 4083.

8.   (a) Posner, G. H.; Nelson, T. D.; Kinter, C. M.; Afarinkia, K. *Tetrahedron Lett.* **1991,** *32,* 5295; (b) For a review of 2-pyrone cycloadditions see: Afarinkia, K.; Vinader, M. V.; Nelson, T. D.; Posner, G. H. *Tetrahedron Report* **1992,** *48,* 9111. (c) For applications to synthesis of vitamin D3 analogs, see Posner, G. H.; Dai, H. *BioMed. Chem. Lett.,* **1993,** *3,* 1829 (d) For cycloadditions using 5-bromo-2-pyrone, see Afarinkia, K.; Posner, G. H. *Tetrahedron Lett.* **1992,** *33,* 7839.

Appendix
Chemical Abstracts Nomenclature (Collective Index Number);
(Registry Number)

3-Bromo-2(H)-pyran-2-one: 2H-Pyran-2-one, 3-bromo- (8,9); (19978-32-6)

3-Bromo-5,6-dihydro-2(H)-pyran-2-one: 2H-Pyran-2-one, 3-bromo-5,6-dihydro- (11); (104184-64-7)

5,6-Dihydro-2(H)-pyran-2-one: 2H-Pyran-2-one, 5,6-dihydro- (8,9); (3393-45-1)

Dichloromethane: Methane, dichloro- (8,9); (75-09-2)

Bromine (8,9); (7726-95-6)

Triethylamine (8); Ethanamine, N,N-diethyl- (9); (121-44-8)

3,5-Dibromo-5,6-dihydro-2(H)-pyran-2-one: 2H-Pyran-2-one, 3,5-dibromo-5,6-dihydro-, (±); (56207-18-2)

N-Bromosuccinimide: Succinimide, N-bromo- (8); 2,5-Pyrrolidinedione,

1-bromo- (9); (128-08-5)

Benzoyl peroxide (8); Peroxide, dibenzoyl (9); (94-36-0)

Carbon tetrachloride: CANCER SUSPECT AGENT (8); Methane, tetrachloro- (9);

(56-23-5)

1,3,5-CYCLOOCTATRIENE

A.

$$\text{NBS/CCl}_4 \atop \text{reflux}$$

B.

$$\text{Li}_2\text{CO}_3\text{-LiCl} \atop \text{DMF, 90-95°C}$$

Submitted by Masaji Oda, Takeshi Kawase, and Hiroyuki Kurata.[1]

Checked by Daniel V. Paone and Amos B. Smith, III.

1. Procedure

A. Allylic bromination of 1,5-cyclooctadiene. A 2-L, three-necked, round-bottomed flask, equipped with a mechanical stirrer, reflux condenser, and a heating mantle, is charged with 216.4 g (2.0 mol) of 1,5-cyclooctadiene (Note 1), 44.5 g (0.25 mol) of N-bromosuccinimide (NBS), 0.5 g of benzoyl peroxide, and 700 mL of carbon tetrachloride. The mixture is heated to gentle reflux with stirring. When the reaction starts, a rapid reflux is observed. Three more 44.5-g portions (0.25 mol) of NBS are added at 30-min intervals (total 178 g, 1.0 mol). Heating is continued for 1.5 hr after addition of the final portion of NBS. The mixture is cooled to room temperature and suction filtered and the filter cake is washed with 150 mL of carbon tetrachloride (Note 2). The filtrate is washed once with 150 mL of water, dried over calcium chloride, and

240

filtered with suction. A vacuum distillation apparatus consisting of a 500-mL, two-necked, round-bottomed flask, a stoppered pressure-equalizing dropping funnel, a distilling head (25 cm long) packed with Pyrex glass tips or helices, a condenser, and receivers is assembled (Note 3). The dried carbon tetrachloride solution is transferred to the dropping funnel, the system is evacuated to 150 mm, and the solution is introduced continuously from the dropping funnel resulting in removal of the bulk of the solvent (Note 4). The pale yellow residue is then fractionally distilled first at 30 mm to remove the unreacted 1,5-cyclooctadiene (Note 5), and then at 5 mm to distill the bromocyclooctadienes to give 113-121 g (60-65%) of a mixture of 3-bromo-1,5-cyclooctadiene and 6-bromo-1,4-cyclooctadiene,[2] bp 66-69°C at 5 mm (Notes 6 - 8).

B. *1,3,5-Cyclooctatriene.* A 1-L, three-necked, round-bottomed flask, equipped with a magnetic stirring bar, pressure-equalizing dropping funnel, immersion thermometer, and a condenser bearing a gas inlet vented through a mineral oil bubbler, is charged with 25.9 g (0.35 mol) of lithium carbonate, 2.0 g (0.047 mol) of lithium chloride (Note 9), and 400 mL of dry dimethylformamide (DMF) (Note 10). The magnetically stirred mixture is heated to 90°C in an oil bath (Note 11) and 113.5 g (0.607 mol) of the bromocyclooctadiene mixture (Part A) is added dropwise via the dropping funnel over 50 min. During the addition, rapid evolution of gas (carbon dioxide) is observed via the bubbler. After completion of the addition, heating is continued for 1 hr at 90-95°C. The mixture is cooled to room temperature, diluted with 1 L of ice water, and the mixture extracted twice with 200-mL portions of pentane. The combined organic phase is washed twice with 100-mL portions of water, dried over sodium sulfate, and filtered. The filtrate is distilled at atmospheric pressure to remove the pentane, and the residue is distilled under reduced pressure, employing a short (12 cm) Vigreux column, to give 54-58 g (84-90%) of almost pure 1,3,5-cyclooctatriene, bp 63-65°C at 48 mm (Note 12).

2. Notes

1. All reagents and solvents are commercially available and are used without further purification.

2. The solids consisted of 94.0 g of succinimide (0.95 mol, 95% of theoretical).

3. The distillation system was connected to a closed-tube manometer and a Cartesian diver-type pressure regulator employed to control the pressure.

4. When a rotary evaporator was used for the concentration, the recovery of 1,5-cyclooctadiene decreased substantially.

5. Approximately 88.5 g (0.82 mol, 82% of theoretical) of 1,5-cyclooctadiene, bp 55-57°C at 30 mm, is recovered and can be recycled.

6. Care must be taken during fractionation of 1,5-cyclooctadiene and the bromocyclooctadienes, because contamination of the bromide with 1,5-cyclooctadiene leads to contamination of 1,3,5-cyclooctatriene with the diene. A 1-2-mL intermediate fraction effects clean separation. The distillation took the checkers 6-8 hr. The bromides are extremely light sensitive, turning yellow to red-brown quickly. To avoid product coloration all product receiving flasks were wrapped in aluminum foil.

7. When NBS was added in two portions instead of four, the yield of bromocyclooctadienes decreased slightly to 60%. A preparation using 500 g (2.80 mol) of NBS (five-portion addition) gave a higher yield (78%).

8. The spectral data for the mixture of bromocyclooctadienes is as follows: ^{1}H NMR (500 MHz, CDCl$_3$) δ: 1.7-3.4 (m, 12 H), 4.6-5.25 (m, 2 H), 5.3-5.9 (m, 8 H); ^{13}C NMR (125 MHz, CDCl$_3$) δ: 25.0, 27.7, 28.3, 28.7, 34.3, 36.8, 48.0, 49.9, 124.7, 127.0, 128.3, 128.8, 129.2, 130.4, 130.6, 131.7; IR (thin film) cm^{-1}: 3000, 2940, 2890, 2820, 1640, 1480, 1450, 1440, 1420, 1220, 1150, 1140, 990, 910, 860, 805, 670.

9. Lithium carbonate and lithium chloride were dried under reduced pressure at 80-100°C for 3 hr before use.

10. DMF dried azeotropically with benzene is sufficient for the present reaction.

11. The checkers employed a heating mantle.

12. 1,3,5-Cyclooctatriene exists in equilibrium with bicyclo[4.2.0]octa-2,4-diene, its valence isomer (ratio = ~7:1).[3] 1,3,5-Cyclooctatriene exhibits the following spectral data: ^{1}H NMR (500 MHz, CDCl$_3$) δ: 2.43 (s, 4 H), 5.50-6.00 (m, 6 H); ^{13}C NMR (125 MHz, CDCl$_3$) δ: 28.0, 125.9, 126.7, 135.5; IR (thin film) cm^{-1}: 3000, 2920, 2875, 2830, 1635, 1605, 1445, 1425, 1220, 690, 635. Signals for the minor valence isomer may be observed. No signals for 1,3,6-cyclooctatriene, another possible isomer, are observed.

Waste Disposal Information

All toxic materials were disposed of in accordance with "Prudent Practices for Disposal of Chemicals from Laboratories"; National Academic Press; Washington, DC, 1983.

3. Discussion

Mixtures of 1,3,5- and 1,3,6-cyclooctatriene were obtained by partial reduction of cyclooctatetraene in ways such as protonation of cyclooctatetraene dianion[3d,4] and reduction with zinc-alkali.[2,5] 1,3,6-Cyclooctatriene is the major product in these reductions. However, since 1,3,6-cyclooctatriene isomerizes to 1,3,5-cyclooctatriene on treatment with base, quenching cyclooctatetraene dianion with methanol and subsequent heating affords 1,3,5-cyclooctatriene in an 80% yield.[3d] Reduction of cyclooctatetraene with sodium hydrazide and hydrazine also produces 1,3,5-cyclooctatriene.[6] Therefore, when cyclooctatetraene is available in quantity, these procedures are the methods of choice.

The present two-step procedure for the synthesis of 1,3,5-cyclooctatriene uses commercially available 1,5-cyclooctadienes as starting material. Although allylic bromination of 1,5-cyclooctadiene with N-bromosuccinimide produces a mixture of 3-bromo-1,5-cyclooctadiene and 6-bromo-1,4-cyclooctadiene,[2] dehydrobromination of this mixture with LiCl-Li_2CO_3/DMF affords only 1,3,5-cyclooctatriene, which is in equilibrium with its valence isomer bicyclo[4.2.0]octa-2,4-diene.

1. Department of Chemistry, Faculty of Science, Osaka University, Toyonaka, Osaka 560, Japan.

2. Echter, T.; Meier, H. *Chem. Ber.* **1985**, *118*, 182.

3. (a) Cope, A. C.; Haven, A. C., Jr.; Ramp. F. L.; Trumbull, E. R. *J. Am. Chem. Soc.* **1952**, *74*, 4867; (b) Kröner, M. *Chem. Ber.* **1967**, *100*, 3172; (c) Huisgen, R.; Boche, G.; Dahmen, A.; Hechtl, W. *Tetrahedron Lett.* **1968**, 5215; (d) Adam, W.; Gretzke, N.; Hasemann, L.; Klug, G.; Peters, E. M.; Peters, K.; von Schnering, H. G.; Will, B. *Chem. Ber.* **1985**, *118*, 3357.

4. Reppe, W.; Schlichting, O.; Klager, K.; Toepel, T. *Justus Liebigs Ann. Chem.* **1948**, *560*, 1; Cope, A. C.; Hochstein, F. A. *J. Am. Chem. Soc.* **1950** *72*, 2515.

5. Jones, W. O. *J. Chem. Soc.* **1954**, 1808; Sanne, W.; Schlichting, O. *Angew. Chem.* **1963**, *75*, 156.

6. Kauffmann, T.; Kosel, C.; Schoeneck, W. *Chem. Ber.* **1963**, *96*, 999.

Appendix

Chemical Abstracts Nomenclature (Collective Index Number); (Registry Number)

1,3,5-Cyclooctatriene (8,9); (1871-52-9)

1,5-Cyclooctadiene (8,9); (111-78-4)

N-Bromosuccinimide: Succinimide, N-bromo- (8); 2,5-Pyrrolidinedione, 1-bromo- (9); (128-08-5)

Benzoyl peroxide (8); Peroxide, dibenzoyl (9); (94-36-0)

Carbon tetrachloride: CANCER SUSPECT AGENT (8); Methane, tetrachloro- (9); (56-23-5)

3-Bromo-1,5-cyclooctadiene: 1,5-Cyclooctadiene, 3-bromo- (8,9); (23346-40-9)

6-Bromo-1,4-cyclooctadiene: 1,4-Cyclooctadiene, 6-bromo- (8,9); (23359-89-9)

N,N-Dimethylformamide: CANCER SUSPECT AGENT: Formamide, N,N-dimethyl- (8,9); (68-12-2)

3-PYRROLINE

(1H-Pyrrole, 2,5-dihydro-)

A.

B.

C.

Submitted by Albert I. Meyers,[1] Joseph S. Warmus, and Garrett J. Dilley.

Checked by Kevin W. Gillman and David J. Hart.

1. Procedure

A. *Preparation of 1-[(Z)-4-chloro-2-butenyl]-1-azonia-3,5,7-triazatricyclo-[3.3.1.1³,⁷]decane chloride* (**2**). A 1-L, single-necked, round-bottomed flask is charged with 34.4 g (246 mmol) of hexamethylenetetramine and 500 mL of chloroform ($CHCl_3$) (Note 1). cis-1,4-Dichlorobut-2-ene (30.2 g, 242 mmol) is added, the flask is fitted with a reflux condenser (Note 2), and the mixture is heated to reflux with a heating mantle.

246

After 4 hr, the mixture is cooled to room temperature, and filtered through a sintered glass funnel (10-20 μ porosity). The resulting white solid, quaternary salt **2**, is washed with two 100-mL portions of CHCl₃, and the filtrate is heated to reflux for an additional 18 hr, cooled, and filtered. The resulting light brown solid **2** is washed twice with 50-mL portions of CHCl₃. The combined solids are dried in a desiccator under reduced pressure (1 mm) to afford 58.6 g (91%) of **2** (Note 3).

B. Preparation of (Z)-4-chloro-2-butenylammonium chloride (**3**). A 1-L, single-necked, round-bottomed flask, equipped for magnetic stirring, is charged with 400 mL of 95% ethanol (EtOH) and 70 mL of concd hydrochloric acid (HCl) is slowly added (slightly exothermic reaction). To the still warm solution is added solid **2** (58.5 g, 221 mmol) in one portion. The reaction initially becomes homogeneous and slightly orange colored, then after 45 min a precipitate begins to form. The reaction mixture is allowed to stir at room temperature for 18 hr, cooled to 0°C, and the precipitate (NH₄Cl) is collected by filtration using a 10-20 μ porosity sintered glass funnel. The collected solid is washed on the funnel twice with 100 mL of cold 95% EtOH. The filtrate is concentrated by rotary evaporation, and the remaining semisolid is taken up in 40 mL of cold 95% EtOH. The resulting precipitate (NH₄Cl) is again collected on the filter and washed with cold 95% EtOH (2 x 20 mL). This procedure (rotary evaporation, solution in 95% EtOH, cooling, filtration) is repeated once more.

The solid remaining upon concentration of the filtrate from the final filtration is dissolved in 80 mL of warm ethyl acetate (EtOAc). Upon cooling, the product crystallizes. Hexane is added (30 mL) and the crystals are collected on a 10-20 μ porosity sintered-glass funnel. The solid is washed with 60 mL of hexane. The filtrate is concentrated by rotary evaporation to give an additional amount of solid that is recrystallized from a minimum amount of EtOAc and hexane as before. The combined solids are dried in a desiccator under reduced pressure (1 mm) for 2 hr to give 28.1-30.4 g (90-96%) of **3** as a light yellow crystalline solid (Note 4).

C. Preparation of 3-pyrroline (**4**). A 200-mL round-bottomed flask, equipped with a reflux condenser, is charged with 66.4 g (437 mmol) of 1,8-diazabicyclo[5.4.0] undec-7-ene (DBU) which is cooled to 0°C in an ice bath while **3** (30.3 g, 213 mmol) is added in portions through the top of the condenser. Toward the end of the addition, the resulting slurry becomes orange, gas is evolved, and mixture begins to reflux (the reaction is exothermic). Any remaining salt **3** is added, and when boiling subsides, a short-path distillation head wrapped in glass wool is put in place (Note 5). A 50-mL receiving flask is totally immersed in a -78°C bath (dry ice-isopropyl alcohol). The orange solid mixture is heated using a heating mantle during which time the solid liquifies and 3-pyrroline (**4**) distills at 85-92°C at atmospheric pressure. (*Caution: some foaming occurs initially, but subsides during the reaction*). Heating is continued until no more 3-pyrroline distills, affording 11.05 g (75%) of 3-pyrroline as a clear oil (Notes 6, 7).

2. Notes

1. Hexamethylenetetramine and cis-1,4-dichloro-2-butene (95%) were obtained from Aldrich Chemical Company, Inc., and were not purified before use. 1,8-Diazabicyclo[5.4.0]undec-7-ene (DBU), chloroform, ethyl acetate, and hexane were distilled prior to use.

2. The mixture is left open to air.

3. All physical data for **2** are consistent with that given in reference 14: mp 160-170°C dec; ^{1}H NMR (D$_2$O) δ: 3.54 (dd, 2 H, J = 7.9, 0.5), 4.10 (dd, 2 H, J = 8.1, 0.5), 4.43 (d, 3 H, J = 12.9), 4.58 (d, 3 H, J = 12.9), 5.01 (s, 6 H), 5.67 (m, 1 H), 6.23 (complex m, 1 H); ^{13}C NMR (D$_2$O) δ: 40.5, 55.4, 73.0, 81.1, 119.6, 140.9. The yield of the first crop is 81-82%. The second crop contains a minor impurity as indicated by a doublet at δ 3.79 in the ^{1}H NMR, but is used in the next step.

4. The physical properties of **3** are as follows: mp 117-120°C dec; ^{1}H NMR (D$_2$O) δ: 3.63 (d, 2 H, J = 7.3), 4.07 (d, 2 H, J = 8.0), 5.58 (m, 1 H), 5.89 (m, 1 H); ^{13}C NMR (D$_2$O) δ: 38.8, 41.4, 127.5, 134.9. This material is easily stored without any special precautions.

5. The distillation should be carried out under an inert atmosphere.

6. The 3-pyrroline (**4**), prepared as described above, is estimated to be >95% pure. A sample was stored at 0°C in a stoppered, round-bottomed flask for six months with very little oxidation or decomposition; the compound, however, had yellowed. For prolonged storage, a sealed ampoule is recommended. The checkers obtained a somewhat lower yield of **4** using an oil bath at 200°C for the distillation, whereas the submitters employed a heat gun and obtained a 63% yield of **4** (bp 80-85°C).

7. The NMR spectra of 3-pyrroline (**4**) are as follows: ^{1}H NMR (CDCl$_3$) δ: 1.93 (s (broad), 1 H), 3.71 (s, 4 H), 5.84 (s (broad), 2 H); ^{13}C NMR (CDCl$_3$) δ: 53.6, 128.2.

Waste Disposal Information

All toxic materials were disposed of in accordance with "Prudent Practices for Disposal of Chemicals from Laboratories"; National Academic Press; Washington, DC, 1983.

3. Discussion

3-Pyrroline is a desirable starting material for alkylation of heterocycles. 1-(Methoxycarbonyl)-3-pyrroline[2] has been used to prepare 2,5-dialkylated pyrrolines,[3] which resulted in the synthesis of the Pharaoh ant trail pheromone[4a] and gephyrotoxin 223.[4b] Alkylation of 3-pyrrolines has also led to the synthesis of 12-azaprostaglandins.[5] The submitters have used a formamidine derived from 3-

pyrroline to provide access to 2-substituted pyrrolines and pyrrolidines,[6] which has led to the synthesis of the unnatural (+)-anisomycin.[7]

Recently, Brown has shown the feasibility of a one-carbon homologation procedure using a chiral non-racemic boronate derived from the hydroboration of 3-pyrroline.[8] Pyrrole-containing nucleosides have been prepared from the pyrroline nucleoside by photodehydrogenation.[9]

Pure 3-pyrroline has been difficult to obtain. Commercially available 3-pyrroline was at one time supplied in 85% purity, the remaining 15% being pyrrolidine.[10a] It is now supplied in only 65% purity. Material of 97% purity is available; however, the cost ($51/g) is excessively high, limiting its use as a starting material.[10]

Preparation of N-alkyl-3-pyrrolines has been accomplished by treatment of cis-1,4-dihalo-2-butene with the appropriate amine.[11] However, synthesis of the parent 3-pyrroline by condensation of cis-1,4-dihalo-2-butene with ammonia is a very low-yielding process.[11b]

Reduction of pyrrole by zinc/hydrochloric acid (Zn/HCl) leads to various amounts of pyrrolidine as overreduced material.[12a,b] Other preparations of pure 3-pyrroline were found to be difficult or of low yield.[12] Separation of 3-pyrroline from pyrrolidine is difficult, as they differ in boiling points by only 1.5°C.[12b] Crystallization of their hydrochloride salts,[12b] or urethanes,[13] is possible, but only with significant losses.

A three-step preparation, based on the Delépine reaction,[14] describes the synthesis of this compound in high purity.[15] However, some difficulties were encountered in the hands of the submitters following this procedure.[16] Several modifications have now led to an efficient preparation of 3-pyrroline in high purity, and to a procedure that is readily amenable to large scale synthesis.

1. Department of Chemistry, Colorado State University, Fort Collins, CO 80523.

2   Armande, J. C. L.; Pandit, U. K. *Tetrahedron Lett.* **1977**, 897.

3. Macdonald, T. L. *J. Org. Chem.* **1980**, *45*, 193.

4. (a) Ritter, F. J.; Rotgans, I. E. M.; Talman, E.; Verwiel, P. E. J.; Stein, F. *Experientia* **1973**, *29*, 530; (b) Daly, J. W.; Brown, G. B.; Mensah-Dwumah, M. *Toxicon* **1978**, *16*, 163.

5. Lapierre Armande, J. C.; Pandit, U. K. *Recl. Trav. Chim. Pays-Bas.* **1980**, *99*, 87.

6. (a) Meyers, A. I.; Dickman, D. A.; Bailey, T. R. *J. Am. Chem. Soc.* **1985**, *107*, 7974; (b) Meyers, A. I.; Edwards, P. D.; Rieker, W. F.; Bailey, T. R. *J. Am. Chem. Soc.* **1984**, *106*, 3270.

7. Meyers, A. I.; Dupre, B. *Heterocycles* **1987**, *25*, 113.

8. (a) Brown, H. C.; Gupta, A. K.; Rangaishenvi, M. V.; Vara Prasad, J. V. N. *Heterocycles* **1989**, *28*, 283; (b) Brown, H. C.; Vara Prasad, J. V. N.; Gupta, A. K. *J. Org. Chem.* **1986**, *51*, 4296.

9. (a) Kawana, M.; Emoto, S. *Bull. Chem. Soc. Jpn.* **1969**, *42*, 3539; (b) Kawana, M.; Emoto, S. *Bull. Chem. Soc. Jpn.* **1968**, *41*, 2552.

10. (a) Aldrich Catalog, p. 1080; (b) Aldrich, Catalog Handbook of Fine Chemicals, 1994-1995, p. 1223.

11. (a) Mahboobi, S.; Fischer, E. C.; Eibler, E.; Wiegrebe, W. *Arch. Pharm.* **1988**, *321*, 423; (b) Bobbitt, J. M.; Amundsen, L. H.; Steiner, R. I. *J. Org. Chem.* **1960**, *25*, 2230; (c) Ding, Z.; Tufariello, J. J. *Synth. Commun.* **1990**, *20*, 227.

12. (a) Andrews, L. H.; McElvain, S. M. *J. Am. Chem. Soc.* **1929**, *51*, 887; (b) Hudson, C. B.; Robertson, A. V. *Tetrahedron Lett.* **1967**, 4015; (c) Carelli, V.; Morlacchi, F. *Ann. Chim. (Rome)* **1964**, *54*, 1291; (d) Appel, R.; Büchner, O. *Angew. Chem.* **1962**, *74*, 430; (e) Zwierzak, A.; Gajda, T. *Tetrahedron Lett.* **1974**, 3383; (f) Palmer, B. D.; Denny, W. A. *Synth. Commun.* **1987**, *17*, 601.

13. Colegate, S. M.; Dorling, P. R.; Huxtable, C. R. *Aust. J. Chem.* **1984**, *37*, 1503.

14. Angyal, S. J. *Org. Reactions* **1954**, *8*, 197.

15. Brandänge, S.; Rodriguez, B. *Synthesis* **1988**, 347.

16. Warmus, J. S.; Dilley, G. J.; Meyers, A. I. *J. Org. Chem.* **1993**, *58*, 270.

Appendix
Chemical Abstracts Nomenclature (Collective Index Number);
(Registry Number)

3-Pyrroline (8); 1H-Pyrrole, 2,5-dihydro- (9); (109-96-6)

1-[(Z)-4-Chloro-2-butenyl]-1-azonia-3,5,7-triazatricyclo[3.3.1.1^{3,7}]decane chloride: 3,5,7-Triaza-1-azoniatricyclo[3.3.1.1^{3,7}]decane, 1-(4-chloro-2-butenyl)-, chloride, (Z)- (12); (117175-09-4)

Hexamethylenetetramine (8); 1,3,5,7-Tetrazatricyclo[3.3.1.1^{3,7}]decane (9); (100-97-0)

cis-1,4-Dichlorobut-2-ene: 2-Butene, 1,4-dichoro-, (Z)- (8,9); (1476-11-5)

(Z)-4-Chloro-2-butenylammonium chloride: 2-Butenylamine, 4-chloro-, hydrochloride, (Z)- (9); 2-Buten-1-amine, 4-chloro-, hydrochloride, (Z)- (12); (7153-66-4)

1,8-Diazabicyclo[5.4.0]undec-7-ene (DBU): Pyrimido[1,2-a]azepine, 2,3,4,6,7,8,9,10-octahydro- (8,9); (6674-22-2)

Chloroform: HIGHLY TOXIC. CANCER SUSPECT AGENT.

2-CYCLOHEXENE-1,4-DIONE

A.

B.

C.

Submitted by Masaji Oda, Takeshi Kawase, Tomoaki Okada, and Tetsuya Enomoto.[1]

Checked by Kazushige Kajita and Amos B. Smith, III.

1. Procedure

A. *endo-Tricyclo[6.2.1.0²,⁷]undeca-4,9-diene-3,6-dione* **1**. A 1-L, three-necked, round-bottomed flask, equipped with a mechanical stirrer, 100-mL pressure-equalizing dropping funnel, and a thermometer, is charged with 108.1 g (1.0 mol) of 1,4-benzoquinone (Note 1) and 350 mL of dichloromethane (Note 2). The flask is cooled to 0°C with an ice bath. Via the addition funnel, 41.5 mL of cyclopentadiene (34.2 g,

0.52 mol) (Note 3) is added dropwise over a 45-min period with stirring at such a rate that the internal temperature remains below 8°C (Note 4). A second 41.5-mL portion (34.2 g, 0.52 mol) of cyclopentadiene is added in a similar way (Note 5). The resulting mixture is stirred for 1 hr in the ice bath, for 0.5 hr at room temperature, and then is transferred to a 1-L, pear-shaped flask by rinsing with a small amount of dichloromethane. Most of the solvent is removed by rotary evaporation under reduced pressure, and 200 mL of hexane is added to the residue. The flask is cooled in an ice bath for 0.5 hr. The pale yellow solids are collected by suction filtration, washed with three 30-mL portions of hexane, and air-dried to afford 150-155 g of the Diels-Alder adduct **1** (D-A adduct) of sufficient purity for use in the next step (Note 6). Concentration of the filtrate to about half the volume and cooling affords 12-15 g of a second crop of **1**. The combined yield of **1** is 164-169 g (94-97%).

 B. *endo-Tricyclo[6.2.1.0^{2,7}]undec-9-ene-3,6-dione* **2**. A 3-L, three-necked, round-bottomed flask, equipped with a mechanical stirrer, thermometer, and heating mantle, is charged with 166 g (0.95 mol) of the crude D-A adduct **1** and 1.3 L of acetic acid (Note 7). Portionwise, 228.8 g (3.5 mol) of zinc dust (Note 8) is added to the mixture with stirring. The temperature rises to about 70°C over a 10-min period and then drops. The resulting mixture is then heated at 70-80°C for 1 hr, followed by addition of a further 32.7 g (0.5 mol) of zinc powder and continued heating and stirring for an additional 1.5 hr. After the mixture has cooled to room temperature, the gray solids are collected by suction filtration, and washed with 200 mL of acetic acid (Note 9). The dark yellowish filtrate is then concentrated by rotary evaporation under reduced pressure (Note 10), the residue diluted with 600 mL of water, and the resulting mixture extracted with three 300-mL portions of toluene (Note 11). The combined toluene extracts are washed with two 100-mL portions of water, 100 mL of aqueous 10% sodium hydroxide solution, two 100-mL portions of brine, and dried over sodium sulfate, and filtered into a 500-mL round-bottomed flask. Removal of the solvent by rotary evaporation under reduced pressure yields 148-155 g of crude diketone **2**, sufficiently pure for further transformation, as a reddish liquid that solidifies on cooling (Note 12).

C. *2-Cyclohexene-1,4-dione* **3** (Notes 13 and 14). The 500-mL, round-bottomed flask containing ~150 g of crude diketone **2** from Part B is equipped with a magnetic stirring bar and fitted with a Claisen-type short path distillation head leading to a condenser and a 200-mL, round-bottomed receiver. Two cold traps are placed between the receiver and a vacuum pump. The first trap is cooled in a dry ice-ethanol bath and the second trap in liquid nitrogen (Note 15). The system is evacuated and the distillation pot is placed in an oil bath and heated rapidly to 140-180°C with magnetic stirring, whereupon diketone **2** distills fairly rapidly into the receiver to afford a mixture of diketone **2** and 2-cyclohexene-1,4-dione **3**, bp 80-140°C (most at 130-140°C) at 4-7 mm (bath temperature 140-180°C), as a pale yellow oil (Notes 16, 17 and 18).

The 200-mL, round-bottomed receiver, containing a mixture of diketone **2** and 2-cyclohexene-1,4-dione **3**, is then fitted with a long path (25 cm) distillation head packed with Pyrex glass helices or chips, an air condenser (25 cm), and a 200-mL, two-necked, round-bottomed receiver with one neck connected to the vacuum pump via the traps described earlier. The distillation pot is placed in an oil bath, the system is evacuated, and the receiver is immersed in an ice bath. The bath is heated to 120°C and the mixture of diketone **2** and 2-cyclohexene-1,4-dione **3** is redistilled at 4-8 mm while the bath temperature is gradually increased from 120°C to 190°C. Early in the distillation, the enedione **3**, which is already present, distills smoothly, then the rate of dissociation of **2** becomes rate-determining. The bath temperature is maintained at 180-190°C so as to keep the boiling point nearly constant (87°C/6 mm). Care is taken to assure that any enedione **3** that crystallizes in the condenser or receiver is quickly melted with an efficient heat gun to avoid occlusion of the distillation path (Note 19). A total of 73-80 g of crude yellow solid enedione **3** is obtained containing a small amount of diketone **2**.

The resulting crude solid enedione **3** is melted at 60-70°C in a water bath and 35 mL of carbon tetrachloride is added (Note 20). The mixture is cooled in an ice bath, 18 mL of hexane is added (Note 21), and the resulting mixture is gently stirred with a glass rod. After cooling for 20-30 min, the resulting pale yellow crystals are collected by suction filtration, washed with 20-30 mL of a cold mixture of carbon tetrachloride and hexane (1:1 v/v), and then with 40 mL of hexane. Brief air drying affords 65-71 g of almost pure enedione **3**, mp 54-54.5°C (Notes 21 and 22). Concentration of the filtrate to about 15 mL, addition of 8 mL of hexane, and cooling the mixture furnishes a further 1.5-2.5 g of **3**. In total, 67.0-72.5 g (61-66% from 1,4-benzoquinone) of nearly pure **3** is obtained (Notes 23, 24 and 25).

2. Notes

1. Reagent-grade 1,4-benzoquinone was purchased from Wako Pure Chemical Industries, LTD and was used without additional purification.

2. Dichloromethane was distilled from phosphorus pentoxide.

3. Cyclopentadiene is prepared according to the procedure in *Org. Synth., Coll. Vol. VII* **1990**, 339.

4. At higher temperatures, 1:2 adducts[2] between 1,4-benzoquinone and cyclopentadiene are formed as by-products. Separation of the 1:1 adduct and 1:2 adducts is not easy on a large scale. If the material is contaminated by a small amount of the 1:2 adduct, it is best to carry the procedure on to the last step (see Note 23).

5. Cyclopentadiene was added in two portions to minimize dimerization during warming to room temperature. If a dropping funnel with cooling jacket is available, it is not necessary to divide the cyclopentadiene into two portions.

6. Recrystallization of a portion of the product from hexane-dichloromethane furnishes pure **1**, mp 78-79°C which exhibits the following spectral data: ^1H NMR (500 MHz, CDCl$_3$) δ: 1.39 (d, 1 H, J = 9), 1.51 (d, 1 H, J = 9), 3.19 (s(br), 2 H), 3.51 (s(br), 2 H), 6.03 (s(br), 2 H), 6.54 (s, 2 H); ^{13}C (125 MHz, CDCl$_3$) δ: 48.31, 48.69, 48.74, 135.26, 142.02, 199.39; IR (CHCl$_3$) cm^{-1}: 3000, 2940, 2870, 1675, 1604, 1450, 1335, 1295, 12.75, 1230, 1120, 1105, 1050, 990, 960, 940, 910, 855.

7. Commercial reagent grade acetic acid was used as purchased.

8. Commercial reagent grade zinc dust was used immediately after opening.

9. The gray solids should be added to water as soon as possible after filtration and washing; otherwise, the solids become hot, probably because of air oxidation.

10. Do not distil to dryness.

11. Diethyl ether may be substituted for toluene.

12. Crude diketone **2** may contain some toluene. Chromatography of a portion of the material on silica gel furnishes pure diketone **2**, mp ~ 22°C which exhibits the following spectra data: ^1H NMR (500 MHz, CDCl$_3$) δ: 1.32 (d, 1 H, J = 8), 1.44 (d, 1 H, J = 8), 2.10-2.30 (m, 2 H), 2.50-2.60 (m, 2 H), 3.18 (s(br), 2 H), 3.42 (d, 2 H, J = 1.8), 6.14 (d, 2 H, J = 1.8); ^{13}C (125 MHz, CDCl$_3$) δ: 37.89, 47.38, 48.68, 51.77, 136.58, 209.48; IR (CHCl$_3$) cm^{-1}: 3000, 2940, 2865, 1710, 1415, 1330, 1250, 1220, 1150, 1100, 1050, 995, 895.

13. It is desirable that the following procedures be completed smoothly in one day to avoid possible isomerization of enedione **3** to hydroquinone. If two days are required, the distillate from the first distillation can be stored without change in a freezer in the dark.

14. All glassware used in these procedures should be free of acids or bases.

15. It is important that the first trap be of sufficient size to accommodate liberated cyclopentadiene.

16. A silicone oil bath was used for the pyrolytic distillations.

17. The ratio of diketone **2** to enedione **3** varies with the precise conditions of distillation with **2** being the major product.

18. Rapid distillation serves to eliminate possible contamination of diketone **2** with trace amounts of acids or bases.

19. Toward the end of the pyrolytic distillation enedione **3** often crystallizes on the walls of the distillation head and condenser.

20. Spectrophotometric grade carbon tetrachloride was used without causing isomerization of enedione **3** to hydroquinone.

21. Enedione **3** crystallizes fairly rapidly; stirring controls the crystal size to some extent. Since enedione **3** sublimes easily, drying under reduced pressure is not appropriate.

22. Pure enedione **3** has the following spectral data: ^{1}H NMR (500 MHz, CDCl$_3$) δ: 2.84 (s, 4 H), 6.66 (s, 2 H); ^{13}C (125 MHz, CDCl$_3$) δ: 36.52, 141.05, 197.21; IR (CHCl$_3$) cm^{-1}: 3020, 2970, 2905, 1690, 1600, 1420, 1370, 1300, 1280, 1225, 1140, 1095, 1000, 985, 935, 845.

23. If D-A adduct **1** contains some 1:2 adducts as impurities, 1,4-benzoquinone is formed by a retro-Diels-Alder reaction during the pyrolytic distillations. In this case, a dark yellow solid of benzoquinone can be seen on the walls of the air condenser, and the distillate has a deeper yellow color. Contamination with a small amount of 1,4-benzoquinone apparently does not interfere with photochemical [2+2] cycloadditions of enedione **3** with alkenes and alkynes, an important application of **3**. Fractional distillation of the benzoquinone-contaminated **3** as described for the second distillation of **3** can remove the benzoquinone with some loss of enedione **3**. The benzoquinone deposits initially as a dark yellow solid on the walls of the distillation head and air condenser during early fractions.

24. Enedione **3** as a crystalline solid can be stored indefinitely below 0°C in a glass container in the dark.

25. Experiments starting from 1.5 mol of 1,4-benzoquinone gave similar results.

Waste Disposal Information

All toxic materials were disposed of in accordance with "Prudent Practices for Disposal of Chemicals from Laboratories"; National Academic Press; Washington, DC, 1983.

3. Discussion

2-Cyclohexene-1,4-dione **3** was first synthesized by careful acid hydrolysis of its monoacetal.[3] Because of the high sensitivity of **3** to acids and bases, this method is not suitable for large-scale preparations. The present method is essentially identical to that reported by Chapman, et al.,[4] with modifications for large-scale and minimal solvent use. Alkyl substituted 2-cyclohexene-1,4-diones can be prepared similarly.[4]

The 1,4-benzoquinone-cyclopentadiene Diels-Alder adduct **2** is well known.[5] The procedure described here is adapted for large-scale preparation. For the zinc reduction, aqueous acetic acid has also been used.[4] The present procedure allows recovery of most of the acetic acid. The entire procedure can be completed in 3-4 days.

The most useful application of **3** is its use in photochemical [2+2] cycloadditions with alkenes and alkynes at the carbon-carbon double bond to afford bicyclo[4.2.0]octane-2,5-diones and bicyclo[4.2.0]oct-7-ene-2,5-diones in good to excellent yield.[6] Since selenium dioxide oxidation of the resulting adducts furnishes the corresponding 3-ene-2,5-diones, diketone **3** can be regarded as a 1,4-benzoquinone equivalent leading to [2+2] cycloadducts at the carbon-carbon double bond.

Some typical applications are shown below.[6,7]

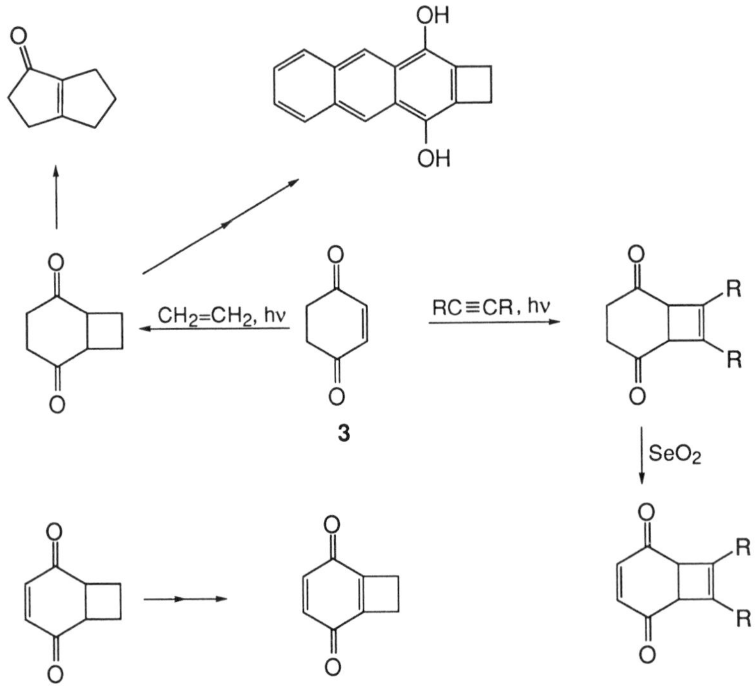

1. · Department of Chemistry, Faculty of Science, Osaka University, Toyonaka, Osaka 560, Japan.

2. Cookson, R. C.; Hill, R. R.; Hudec, J. *J. Chem. Soc.* **1964**, 3043.

3. Garbisch, E. W., Jr. *J. Am. Chem. Soc.* **1965**, *87*, 4971.

4. Chapman, D. D.; Musliner, W. J.; Gates, J. W., Jr. *J. Chem. Soc. C* **1969**, 124.

5. Albrecht, W. *Justus Liebigs Ann. Chem.* **1906**, *348*, 31; Diels, O.; Alder, K. *Justus Liebigs Ann. Chem.* **1928**, *460*, 98; Alder, K.; Flock, F. H.; Beumling, H. *Chem. Ber.* **1960**, *93*, 1896.

6. Oda, M.; Oikawa, H.; Kanao, Y.; Yamamuro, A. *Tetrahedron Lett.* **1978**, 4905; Kawase, T.; Ohnishi, Y.; Oda, M. *J. Chem. Soc., Chem. Commun.* **1991**, 702.

7. Oda, M.; Oikawa, H. *Tetrahedron Lett.* **1980**, *21*, 107; Oda, M.; Kanao, Y. *Chemistry Lett.* **1981**, 37; Sasaki, K.; Kushida, T.; Iyoda, M.; Oda, M. *Tetrahedron Lett.* **1982**, *23*, 2117; Kanao, Y.; Iyoda, M.; Oda, M. *Tetrahedron Lett.* **1983**, *24*, 1727; Watabe, T.; Takahashi, Y.; Oda, M. *Tetrahedron Lett.* **1983**, *24*, 5623.

Appendix
Chemical Abstracts Nomenclature (Collective Index Number); (Registry Number)

2-Cyclohexene-1,4-dione (8,9); (4505-38-8)

endo-Tricyclo[6.2.1.0^{2,7}]undeca-4,9-diene-3,6-dione: 1,4-Methanonaphthalene-5,8-dione, 1,4,4a,8a-tetrahydro- (8,9); (1200-89-1)

endo-Tricyclo[6.2.1.0^{2,7}]undec-9-ene-3,6-dione: 1,4-Methanonaphthalene-5,8-dione, 1,4,4a,6,7,8a-hexahydro- (8,9); (21428-54-6)

p-Benzoquinone (8); 2,5-Cyclohexadiene-1,4-dione (9); (106-51-4)

Cyclopentadiene: 1,3-Cyclopentadiene (8,9); (542-92-7)

CONVERSION OF METHYL KETONES INTO TERMINAL ACETYLENES:
ETHYNYLFERROCENE

(Ferrocene ethynyl)

Submitted by Johann Polin and Herwig Schottenberger.[1]

Checked by Bruce Anderson and Stephen F. Martin.

1.   Procedure

(2-Formyl-1-chlorovinyl)ferrocene: A dry, 1-L, three-necked, round-bottomed flask, equipped with a magnetic stirring bar, an inlet valve for inert gas, a pressure-equalizing addition funnel, and an outlet valve vented through a mercury bubbler, is charged with 22.8 g (0.1 mol) of acetylferrocene (Note 1) and 25 mL (0.32 mol) of N,N-dimethylformamide (DMF) (Note 2). The system is flushed with argon, cooled to 0°C by means of an ice bath, and the brown reaction mixture is stirred well for several minutes (Note 3). Separately, a dry, 100-mL, graduated cylinder bearing a standard taper ground joint with an argon inlet/outlet, is purged with nitrogen and charged with 25 mL (0.32 mol) of DMF. The DMF is cooled in crushed ice and agitated by hand during the cautious addition of 25 mL (0.27 mol) of phosphorus oxychloride (Note 4). The resulting viscous, red complex is transferred to the dropping funnel and added to the magnetically stirred mixture of acetylferrocene and DMF dropwise over 30 min (Note 5). Complete addition is assured by washing the addition funnel and walls of the flask with a small amount of DMF using a pipette. The mixture is stirred at 0°C for 2 hr during

which time the color of the reaction mixture changes from dark brown to olive and ultimately to deep blue. Prior to neutralization, the dropping funnel is replaced by a reflux condenser (Note 6). A 75-mL portion of diethyl ether is added, and the viscous mixture is stirred vigorously for several minutes (Note 7).

Under a positive pressure of argon with continued ice cooling, 116 g (0.85 mol) of sodium acetate trihydrate is cautiously added to the reaction mixture in one portion through a powder funnel followed by cautious addition of 10 mL of water with vigorous stirring (Note 8). The ice bath is removed whereupon the organic layer undergoes a striking color change from colorless to ruby red indicating the formation of the formyl derivative. After 1 hr, an additional 10 mL of ether is added, and stirring is continued for 3 hr at room temperature to ensure complete quenching. The reaction mixture is transferred to a 2-L separatory funnel with ether and water and mixed thoroughly, and the organic phase is separated. The aqueous phase is extracted several times with 100-ml portions of ether (Note 9). The combined organic phases are carefully washed twice with 100-mL portions of saturated aqueous sodium bicarbonate solution (*Caution: gas evolution*) and then with 100 mL of water (Note 10). The organic phase is dried over sodium sulfate, filtered, and concentrated using a rotary evaporator affording 23.4-25.6 g (85-93%) of (2-formyl-1-chlorovinyl)ferrocene (homogeneous by TLC analysis) as deep purple crystals (mp 76-77°C) after drying under high vacuum (Note 11).

Ethynylferrocene: A dry,1-L, three-necked, round-bottomed flask, equipped with a magnetic stirring bar, reflux condenser, and inlet/outlet valves for maintenance of an inert atmosphere as described above, is flushed with argon, charged with 26.0 g (95.0 mmol) of (2-formyl-1-chlorovinyl)ferrocene and 300 mL of anhydrous 1,4-dioxane (Note 12), and the apparatus is placed in an oil bath. The reaction mixture is heated to reflux and after 5 min at reflux, 250 mL of a boiling 1 N solution of sodium hydroxide (a 2.5-fold excess) is cautiously added as rapidly as possible in one portion (Note 13),

263

and the mixture is heated at reflux for another 25 min (Note 14). The oil bath is removed and the reaction mixture is allowed to cool to room temperature.

The reaction mixture is cautiously poured into ice and neutralized with 1 N hydrochloric acid. After transfer to a 1-L separatory funnel, the aqueous mixture is extracted five times with 100 mL of hexane (Note 15). After the combined organic extracts are successively washed twice with 100-mL portions of saturated aqueous sodium bicarbonate solution and water, the organic phase is dried over sodium sulfate, filtered, and concentrated using a rotary evaporator affording an orange residue of crude ethynylferrocene. The crude product is purified by flash chromatography (Silica G-60, 5 x 15 cm column) with elution by hexane (Note 16). Concentration of the fractions containing the product and drying under high vacuum affords 14.8-15.0 g (74 - 75%) of pure ethynylferrocene which crystallizes as an orange solid, mp 53°C (lit.[2] 52-53.5°C) upon seeding (Note 17).

2. Notes

1. Acetylferrocene, $C_{12}H_{12}FeO$, 95% (FW (228.07), mp 85-86°C) is available from Aldrich Chemical Company, Inc. or Lancaster Synthesis Ltd., and is used without further purification (*Caution: highly toxic*). Acetylferrocene should be carefully triturated before use. The analytical data are as follows: ^{1}H NMR (100 MHz, CCl_4) δ: 2.25 (s, 3 H), 4.08 (s, 5 H), 4.30 (s, 2 H), 4.61 (s, 2 H); ^{13}C NMR (22.6 MHz) δ: 26.9 , 69.2, 69.5, 71.8, 79.3, 200.1. IR (CCl_4) cm[-1]: 3100, 1675.

2. N,N-Dimethylformamide (DMF), 99% (C_3H_7NO, FW (73.10), mp -61°C, bp 153°C, d = 0.944, n_D^{20} 1.4305) was purchased from Fluka Chemie AG, and used without further purification. *Caution: DMF is a cancer suspect agent.*

3. The best results were obtained in an argon atmosphere, although from the stability of the product it seems most likely that an inert gas atmosphere is not

essential. Care must be taken to stir the entire system, particularly for large-scale syntheses.

4. Phosphorus oxychloride ($POCl_3$), 99% (FW 153.33, mp 1.25°C, bp 105.8°C, d = 1.645) available from Fluka Chemie AG, was used as purchased. $POCl_3$ is highly toxic and moisture sensitive.

5. *Caution: The formation of the complex is highly exothermic! Be aware of the hazards of phosphorus oxychloride.*

6. This is a safety measure in case the neutralization should become too exothermic.

7. If the ethereal layer turns orange, it is removed and replaced with 75 mL of fresh ether. This procedure removes any traces of unreacted acetylferrocene or ferrocene impurities. The use of a pipette is recommended to replace the organic layer, if necessary.

8. Sodium acetate trihydrate ($CH_3CO_2Na \cdot 3\ H_2O$) 99%, available from Fluka Chemie AG, was used. Anhydrous sodium acetate (CH_3CO_2Na) 99% is only appropriate, if sufficient amounts of water are present.

9. Initially the phase separation is hard to discern. Extraction is continued until the organic phase is nearly colorless.

10. Additional sodium acetate trihydrate is added to the combined aqueous phases, which after some time, affords a small amount of additional product upon ether extraction. Careful extraction and washing of the organic phases prevents undesired polymerization. The yield and quality of the product obtained are largely dependent on the care taken in the extraction procedure.

11. (2-Formyl-1-chlorovinyl)ferrocene ($C_{13}H_{11}ClFeO$) has the following spectroscopic characteristics: ^1H NMR (300 MHz, $CDCl_3$) δ: 4.24 (s, 5 H), 4.57 (s, 2 H), 4.75 (s, 2 H), 6.40 (d, 1 H, J = 6.7), 10.09 (d, 1 H, J = 6.7); IR (CCl_4) cm^{-1}: 2851, 1671.

12. 1,4-Dioxane, 99% ($C_4H_8O_2$, FW (88.11), mp 11.8°C, bp 100-102°C, n_D^{20} 1.4225, d = 1.034) was purchased from Aldrich Chemical Company, Inc. and distilled from sodium benzophenone ketyl before use. *Caution: dioxane is a cancer suspect agent and a flammable liquid.* Attempts to use other solvents failed, and despite subsequent addition of aqueous sodium hydroxide, prior distillation of the dioxane from sodium benzophenone ketyl seems to be essential.

13. A solution of aqueous sodium hydroxide is prepared by dissolving 10 g of sodium hydroxide pellets, 97% (Fluka Chemie AG) in 250 mL of water. The solution is heated to boiling before addition.

14. After this time, TLC analysis (hexane as eluent) indicates essentially complete conversion to ethynylferrocene and an impurity with an R_f value near zero.

15. A pH of 6-7 should be maintained. The phase boundary of the organic and aqueous phases is often difficult to discern, but separation is most satisfactory if the organic/aqueous mixture is filtered through a pad of Celite to remove an oily third phase prior to separation of the aqueous and organic layers. After the final extraction, the organic layer should be nearly colorless.

16. Silica Gel 60741 from Fluka Chemie AG was used. The impurities remain at the top of the 5 x 15-cm column when hexane is used as eluent.

17. The spectroscopic data for ethynylferrocene are as follows: ^1H NMR (300 MHz, $CDCl_3$) δ: 2.71 (s, 1 H), 4.19 (m, 2 H), 4.21 (s, 5 H), 4.46 (m, 2 H); ^{13}C NMR (75 MHz) δ: 63.5, 68.3, 69.6, 71.2; IR (CCl_4) cm^{-1}: 3311, 2112.

Waste Disposal Information

All toxic materials were disposed of in accordance with "Prudent Practices for Disposal of Chemicals from Laboratories"; National Academic Press; Washington, DC, 1983.

3. Discussion

The two-step synthesis of ethynylferrocene described here follows essentially the scale-sensitive route reported by Rosenblum, et al.[2] Although various intermediates have been evaluated,[3,4] (2-formyl-1-chlorovinyl)ferrocene is the most successful precursor in the synthesis of ethynylferrocene. Treatment of acetyl-ferrocene with phosphorus oxychloride in dimethylformamide leads to mixtures of (2-formyl-1-chlorovinyl)ferrocene and the more unstable (1-chlorovinyl)ferrocene, with the ratio of products depending on the stoichiometry.[2] However, production of (1-chlorovinyl)ferrocene can be effectively suppressed by employing an excess of phosphorus oxychloride. Using dimethylformamide as solvent leads to satisfactory results only for small-scale preparations. However, modification of the stoichiometry and experimental conditions led to the above described procedure which is useful for large-scale preparations. Use of conditions employing a comparatively small excess of dimethylformamide and phosphorus oxychloride resulting in a heterogeneous reaction mixture, as well as use of solid sodium acetate trihydrate surmount the problems of scale up and enable the removal of organic impurities. The purity and yield of the intermediate (2-formyl-1-chlorovinyl)ferrocene are substantially improved using the present procedure, and this intermediate is obtained in pure form without need of chromatography .

The procedure for the final elimination reaction is essentially that of Rosenblum, et al.[2] A more detailed procedure is provided which improves reproducibility. Treatment of an ethereal solution of (2-formyl-1-chlorovinyl)ferrocene with sodium amide in liquid ammonia under anhydrous conditions is also an acceptable method,[5] along with the method described which employs base-induced elimination using aqueous sodium hydroxide in dioxane.[2,6] Compounds of the α-haloferrocene type are converted more or less quantitatively into alkynes by dehydrochlorination using

267

potassium tert-butoxide in dimethyl sulfoxide.[7] This alternative method for converting the β-chloroaldehyde might also be useful, but lower yields (15-20% less) make the conventional method[2] more efficient for the synthesis of ethynylferrocene.

With respect to cost and ease of accessibility, the procedure described above is superior to other, more recent synthetic methods.[8-12] However, the most convenient alternative synthesis of ethynylferrocene is that of Doisneau, et al.[11] Some procedures[11,12] also permit the synthesis of 1,1´-diethynylferrocene derivatives. Diethynylmetallocenes represent versatile precursors for the preparation of oligometallocenes.

1. Institut für Allgemeine, Anorganische und Theoretische Chemie Universität Innsbruck, Innrain 52a, A-6020 Innsbruck, Austria.

2. Rosenblum, M.; Brawn, N.; Papenmeier, J.; Applebaum, M. *J. Organomet. Chem.* **1966**, *6*, 173.

3. Benkeser, R. A.; Fitzgerald, Jr., W. P. J. *J. Org. Chem.* **1961**, *26*, 4179.

4. Schlögl, K.; Egger, H. *Monatsh.* **1963**, *94*, 376.

5. Schlögl, K.; Steyrer, W. *Monatsh.* **1965**, *96*, 1520.

6. Bodendorf, K.; Kloss, P. *Angew. Chem.* **1963**, *75*, 139.

7. Abram, T. S.; Watts, W. E. *Synth. React. Inorg. Met.-Org. Chem.* **1976**, *6*, 31.

8. Tsuji, T.; Watanabe, Y.; Mukaiyama, T. *Chem. Lett.* **1979**, 481. This method is not useful for the preparation of ethynylferrocene (expensive; poor yields).

9. Negishi, E.-i.; King, A. O.; Tour, J. M. *Org. Synth., Coll. Vol. VII* **1990**, 63.

10. Pudelski, J. K.; Callstrom, M. R. *Organometallics* **1992**, *11*, 2757.

11. Doisneau, G.; Balavoine, G.; Fillebeen-Khan, T. *J. Organomet. Chem.* **1992**, *425*, 113.

12. Buchmeiser, M.; Schottenberger, H. *J. Organomet. Chem.* **1992**, *441*, 457.

Appendix

Chemical Abstracts Nomenclature (Collective Index Number); (Registry Number)

Ethynylferrocene: Ferrocene, ethynyl- (8,9); (1271-47-2)

(2-Formyl-1-chlorovinyl)ferrocene: Iron, [(1-chloro-2-formylvinyl)cyclo-pentadienyl]cyclopentadienyl- (9); (12085-68-6)

Acetylferrocene: HIGHLY TOXIC: Ferrocene, acetyl- (9); (1271-55-2)

N,N-Dimethylformamide: CANCER SUSPECT AGENT: Formamide, N,N-dimethyl- (8,9); (68-12-2)

Phosphorus oxychloride: HIGHLY TOXIC: Phosphoryl chloride, (8,9); (10025-87-3)

p-Dioxane: CANCER SUSPECT AGENT: (8); 1,4-Dioxane (9); (123-91-1)

4,5-DIBENZOYL-1,3-DITHIOLE-1-THIONE

(Benzenecarbothioic acid, S,S'-(2-thioxo-1,3-dithiole-4,5-diyl) ester)

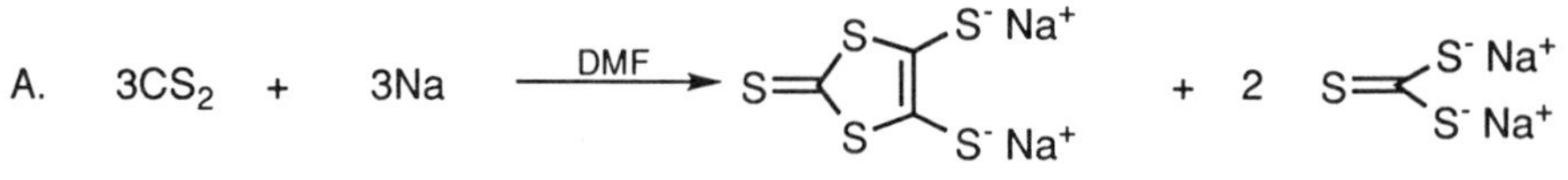

A. $3CS_2 + 3Na \xrightarrow{DMF}$ [structure **1**] $+ 2$ [structure]

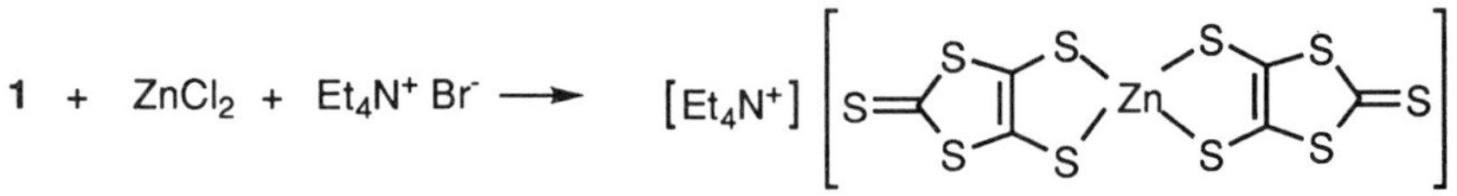

1

$1 + ZnCl_2 + Et_4N^+ Br^- \longrightarrow [Et_4N^+]$ [structure **2**]

2

B. $2 + 4 PhCOCl \longrightarrow$ [structure **3**]

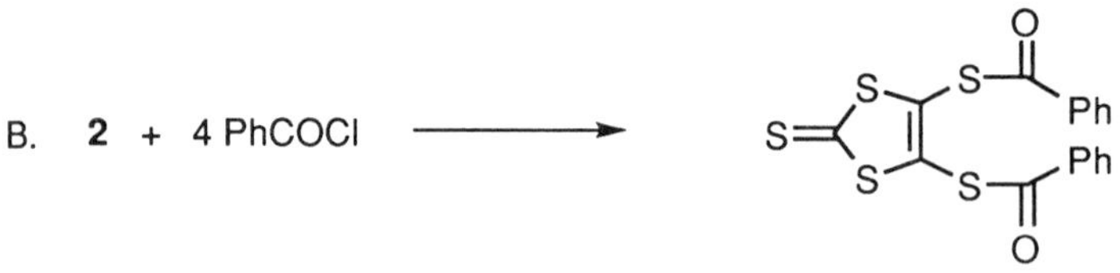

3

Submitted by Thomas K. Hansen,[1] Jan Becher,[1] Tine Jørgensen,[1] K. Sukumar Varma,[2] Rajesh Khedekar,[3] and Michael P. Cava.[3]

Checked by John Hynes and Amos B. Smith, III.

1. Procedure

Caution! The preparation should be carried out in an efficient hood, since carbon disulfide is both toxic and easily ignited.

A. Tetraethylammonium bis(1,3-dithiole-2-thione-4,5-dithiol) zincate (2). An oven-dried (105°C for 6 hr) apparatus consisting of a 3-L, round-bottomed flask, equipped with a mechanical stirrer (Note 1), a 250-mL pressure-equalizing dropping funnel, and a gas inlet tube, is connected to nitrogen (N_2) via a Firestone valve (Note 2). To avoid water condensing on the surfaces, N_2 should be allowed to flow through the flask and out of the dropping funnel while the glassware is still hot. While the dropping funnel is temporarily removed, the flask is charged with 23.0 g of sodium shavings (1.0 mol) (Note 3) via a solid addition funnel (Notes 4 and 5) and, after reassembly, the apparatus is flushed with N_2 for 5 min and placed in an ice-water bath. Carbon disulfide (CS_2, 180 mL, 3.0 mol) (Note 6) is introduced into the flask through the dropping funnel, after which the dropping funnel is immediately loaded with 200 mL of dimethylformamide (DMF) (Note 7), the system is closed, and the N_2 atmosphere is maintained throughout the reaction by adjusting the Firestone valve to a slightly positive pressure (Note 8). The DMF is added dropwise with stirring over 4 hr during which a red color appears (after addition of several mL of DMF). During the addition, the temperature of the reaction mixture is maintained with ice-water cooling. Upon completion of the addition, the reaction mixture is allowed to warm to room temperature and stir overnight, during which time it becomes increasingly deep red/violet. The reaction mixture is visually inspected for residual sodium (Note 9). As a precaution, ice is added to the cooling bath and 50 mL of methanol is added slowly through the dropping funnel (Note 9). A mixture of 400 mL methanol and 500 mL of deionized water, which is degassed by applying vacuum, is then added rapidly through the dropping funnel. A solution of 20 g (0.15 mol) of zinc chloride ($ZnCl_2$) in a mixture of 500 mL of concentrated aqueous ammonium hydroxide and 500 mL of methanol is then added through the dropping funnel (Note 10). A solution of 53 g (0.25 mol) of tetraethylammonium bromide in 250 mL of deionized water is added dropwise via the dropping funnel with vigorous stirring over at least 4 hr, and the solution is stirred overnight. A large amount of red precipitate develops in the flask overnight. This salt collected by suction on a Büchner funnel and washed immediately with 500 mL of deionized water, then with 400 mL portions of isopropyl alcohol until the filtrate is colorless, and finally once with 200 mL of diethyl ether. The product is then dried in a desiccator under vacuum affording 74-76 g (83-84%) of zincate **2** as a red powder. The product is sufficiently pure for further reactions including alkylation and acylation (Notes 11 and 12).

B. *4,5-Dibenzoylthio-1,3-dithiole-1-thione* (**3**). A 1-L, round-bottomed flask, equipped for magnetic stirring, is charged with 400 mL of acetone (Note 13) and 16 g (0.0223 mol) of tetraethylammonium bis(1,3-dithiole-2-thione-4,5-dithio) zincate (**2**). Magnetic stirring is initiated, 40 mL (48.4 g, 0.345 mol) of benzoyl chloride (Note 14) is added *via* a dropping funnel over 4 hr, and the mixture is left overnight. The resulting yellow-light brown precipitate is collected by suction and washed on the filter with 500 mL of deionized water and 300 mL of acetone. This crude material is dissolved in 350 mL of chloroform, 0.5 g of Norit is added and the mixture heated under reflux for 10 min (Note 15). The mixture is filtered while still hot, and the filter cake washed with an additional 50 mL of hot chloroform to dissolve any residual product on the filter. The combined chloroform solutions are concentrated to 150 mL by use of a rotary evaporator. The resulting mixture is warmed to effect complete solution of diester **3** and 50 mL of methanol is added portionwise with stirring. The solution is then left overnight in the refrigerator. The resulting crystalline precipitate is collected by suction and air-dried, affording 11.8-12.0 g (65-66%) of dibenzoyl ester **3** (mp 143-144°C) (Note 16)

2. Notes

1. The mechanical stirrer must be made of inert material (glass or Teflon).

2. If a Firestone valve (Ace Glass Inc., Vineland NJ) is not available, a bubbler connected to the N_2 inlet tube via a T-adapter is sufficient.

3. Sodium shavings (23.0 g, 1 mol) are conveniently prepared in the following way: A clean piece of sodium, 40-60 g, is weighed. Shavings are prepared from this using a commercial household grater directly into a beaker containing dry ether. The weight of shavings is determined by differential weighing of the remaining piece of sodium. *Caution: Sodium is a corrosive and potentially flammable solid. Handle with care using appropriate gloves for protection in a hood avoiding contact with moisture.*

4. *Extreme caution is necessary during the preparation and transfer of the sodium shavings because of the large amount of sodium placed in the flask.* The flask must be free of flaws and sodium waste must be carefully destroyed with ethanol.

5. Decanting the ether off immediately before addition is recommended to avoid oxidation of the shavings. If the shavings are a little wet from ether this does not harm the reaction.

6. The CS_2 was HPLC grade obtained from Aldrich Chemical Company, Inc. and used without purification. Lower grades of CS_2 can be used; however, a minor decrease in yield will occur.

7. The DMF used was obtained from Fisher Scientific Company (certified A.C.S.) and used without further purification.

8. It is important that the stirrer assembly be nearly gas tight, so that only small amounts of N_2 and CS_2 are lost.

9. Normally there will not be any sodium metal left, but small pieces of sodium may have adhered to the walls of the flask above the liquid level. These pieces of residual sodium metal must be destroyed with methanol. If any sodium is left, gas evolution is seen and the mixture is left for 1 hr with cooling and stirring.

10. This is most easily accomplished by adding the $ZnCl_2$ in portions to the rapidly stirred ammonium hydroxide followed by addition of the methanol.

11. Recrystallization of the zinc complex can be carried out as follows: Dissolve 20 g of the salt in 300 mL of warm acetone and add 0.5 g of Norit. Heat at reflux for 10

min, filter the hot mixture to remove the Norit, concentrate the filtrate under reduced pressure to one half the original volume, and add 100 mL of isopropyl alcohol. Collect the resulting precipitate by suction and wash the precipitate with ether. Provides 15.8 g (~80% recovery) of pure salt (mp 206-208°C).

12. The product exhibits the following properties: IR (KBr) cm^{-1}: 1460, 1410, 1165, 1050, 986, 878, 775, 450; ^{13}C NMR (125 MHz, CDCl$_3$) δ: 7.9, 53.2, 136.3, 209.5.

13. Commercial acetone was first dried with CaCl$_2$, then distilled.

14. Benzoyl chloride was obtained from Merck & Company, Inc., and used without purification.

15. Any brand of decolorizing charcoal may be used.

16. The product exhibits the following properties: IR (CHCl$_3$) cm^{-1}: 3001, 1695, 1601, 1586, 1450, 1200, 1179, 1068, 880, 660, 637; ^{13}C NMR (62.9 MHz, CDCl$_3$) δ: 127.8, 129.0, 133.5, 134.6, 134.8, 185.2, 212.1.

Waste Disposal Information

All toxic materials were disposed of in accordance with "Prudent Practices for Disposal of Chemicals from Laboratories"; National Academic Press; Washington, DC, 1983.

3. Discussion

This synthesis provides easy and inexpensive access to the 4,5-dithio-1,3-dithiole-2-thione system. By alkylation with 1,2-dibromoethane, 4,5-ethylenedithio-1,3-dithiole-2-thione, a key intermediate in the synthesis of bis(ethylenedithio)tetrathiofulvalene (BEDT-TTF) is obtained.[4,5] BEDT-TTF forms charge-transfer salts, many of which are conducting and superconducting, and known as "organic metals". Extensive reviews on the synthesis and properties of BEDT-TTF[6] and other TTF derivatives have been published.[7]

Originally, the reduction of carbon disulfide was carried out electrochemically, but this procedure is not practical for preparative amounts of material.[8] In 1979, a chemical reduction method was discovered by Steimecke.[9] The present procedure is based on that 1979 report, but sodium is substituted for potassium (which presents an explosive hazard).[10] Practical improvements have allowed the development of a one-pot procedure that is easily scaled up to provide 70-80 g of material.

Complexation with zinc effectively separates the C_3S_5 dianion from the trithiocarbonate ion in high yield, and the material obtained is sufficiently pure for further reactions. The quaternary ammonium zincate salt is alkylated slowly but smoothly by many halides at room temperature in solvents such as acetone or tetrahydrofuran. Reaction of the zincate with benzoyl chloride followed by cleavage of the resulting benzoate by sodium ethoxide in ethanol provides the much more reactive species $Na_2C_3S_5$. The sodium salt is, however, very air sensitive,[5] whereas the zinc salt is completely stable.

An alternative to this synthesis is provided by Shumaker via tetrathiapentalene.[11] For most purposes, this procedure offers no advantages and requires more expensive starting materials. The procedure described here is applicable not only for the preparation of BEDT-TTF, but intermediates **2** and **3** may also be used for the synthesis of a variety of BEDT-TTF analogs.[12]

1.	Department of Chemistry, Odense University, Campusvej 55, DK-5230, Odense M., Denmark.

2.	Pilkington Technology Centre, Hall Lane Latham Ormskirk, L40 5UF Lancashire, England.

3. Department of Chemistry, The University of Alabama, Box 870336, Tuscalossa, AL 35487-0036.

4. Mizuno, M.; Garito, A. F.; Cava, M. P. *J. Chem. Soc., Chem. Commun.* **1978**, 18-19.

5. Varma, K. S.; Bury, A.; Harris, N. F.; Underhill, A. E. *Synthesis* **1987**, 837-838.

6. Williams, J. M. *Prog. Inorg. Chem.* **1985**, *33*, 183-220.

7. Krief, A. *Tetrahedron* **1986**, *42*, 1209-1252.

8. Wawzonek, S.; Heilmann, S. M. *J. Org. Chem.* **1974**, *39*, 511-514.

9. Steimecke, G.; Sieler, H. J.; Kirmse, R.; Hoyer, E. *Phosphorus Sulfur* **1979**, *7*, 49-55.

10. Moradpour, A.; Schumaker, R. R. *J. Chem. Educ.* **1986**, *63*, 1016.

11. Schumaker, R. R.; Engler, E. M. *J. Am. Chem. Soc.* **1977**, *99*, 5521-5522.

12. (a) Becher, J.; Hansen, T. K.; Malhotra, N.; Bojesen, G.; Bøwadt, S.; Varma, K. S.; Girmay, B.; Kilburn, J. D.; Underhill, A. E. *J. Chem. Soc., Perkin Trans I* **1990**, 175-177; (b) Kini, A. E. M.; Beno, M. A.; Williams, J. M. *J. Chem. Soc., Chem. Commun.* **1987**, 335-336.

Appendix
Chemical Abstracts Nomenclature (Collective Index Number);
(Registry Number)

4,5-Dibenzoyl-1,3-dithiole-1-thione: Benzenecarbothioic acid, S,S'-(2-thioxo-1,3-dithiole-4,5-diyl) ester (10); (68494-08-6)

Carbon disulfide (8,9); (75-15-0)

Tetraethylammonium bis(1,3-dithiole-2-thione-4,5-dithiol) zincate: Ethanaminium, N,N,N-triethyl-, (I-4)-bis[4,5-dimercapto-1,3-dithiole-2-thionato)-S^4,S^5]zincate (2-) (2:1) (10); (72022-68-5)

Sodium (8,9); (7440-23-5)

N,N-Dimethylformamide: CANCER SUSPECT AGENT: Formamide, N,N-dimethyl- (8,9); (68-12-2)

Zinc chloride (8,9); (7646-85-7)

Ammonium hydroxide (8,9); (1336-21-6)

Tetraethylammonium bromide: Ammonium, tetraethyl-, bromide (8); Ethanaminium, N,N,N-triethyl-, bromide (9); (71-91-0)

Benzoyl chloride (8,9); (98-88-4)

Chloroform: HIGHLY TOXIC CANCER SUSPECT AGENT (8); Methane, trichloro- (9); (67-66-3)

AZODICARBOXYLATES: METHYL, ETHYL, t-BUTYL AND BIS(2,2,2-TRICHLOROETHYL)

WARNING

It has been learned that a sample of methyl azodicarboxylate [*Org. Synth., Coll. Vol. IV* **1963**, 411] violently decomposed during a distillation using an electrically heated mantle as a source of heat.

It is again strongly recommended that distillation of methyl or ethyl azodicarboxylate be carried out from a thermally controlled bath, *not* an electrically heated mantle, for the latter may overheat the material being distilled. The bath should have a thermometer in it to keep track of the bath temperature, which should not be allowed to go higher than 130°C. The bath should be lowered at the end of the distillation. The distillation should be shielded on all sides by fixed shielding, as described in "Prudent Practices for Handling Hazardous Chemicals in Laboratories", National Academy Press, Washington, DC (1981), pp. 170-171; see also *Org. Synth., Coll. Vol. VIII* **1993**, p. vi.

Tests show that either methyl or ethyl azodicarboxylate can be detonated by shock or heat (C. S. Sheppard, H. N. Schack, and O. L. Mageli, U.S. Patent 3, 347, 845 [1967]). Methyl azodicarboxylate is far more easily detonated than the ethyl ester. Hence, since the chemical properties of the two esters are similar, ethyl azodicarboxylate is almost always the preferred reagent.

Sheppard, Schack and Mageli have shown that the presence of a shock-stable solvent, e.g., methylene chloride or benzene in the proportion by weight of 30% solvent to 70% methyl or ethyl azodicarboxylate, gives solutions that are stable to shock or heat. It is recommended that either of these azodicarboxylates be stored only as such a solution.

It has been suggested that, as an additional safety measure, methyl and ethyl azodicarboxylates be distilled at 0.1-0.5 mm with a maximum oil bath temperature of 80°C. (S. C. Blackstock, Vanderbilt Department of Chemistry). Blackstock reports b.p. 35-36°C/0.1 mm for dimethyl azodicarboxylate. Extreme care should be taken to maintain the pressure at or below that specified throughout the distillation. The receiver should be chilled in an ice bath. In addiiton, the drying step should be carried out very carefully since residual moisture will lead to hydrolysis during the distillation giving rise to diimide and carbon dioxide.

We know of no instances of explosions with t-butyl azodicarboxylate [*Org. Synth., Coll. Vol. V.* **1973**, 160] or bis(2,2,2-trichloroethyl) azodicarboxylate [*Org. Synth., Coll. Vol. VII* **1990**, 56], which would be expected to be less prone to explosion than the methyl or ethyl azodicarboxylates because of their higher molecular weights. Nevertheless, for safety they too should be prepared and handled behind good shielding and, if they are to be kept for a long time, stored in a shock-stable solvent as described above.

Accepted for checking during the period August 1, 1993

through August 1, 1994. An asterisk (*) indicates that

the procedure has been subsequently checked.

Previously, *Organic Syntheses* has supplied these procedures upon request. However, because of the potential liability associated with procedures which have not been tested, we shall continue to list such procedures but requests for them should be directed to the submitters listed.

2686R	3-Ethenyl-4-methoxycyclobutene-1,2-dione and 2-n-Butyl-6-ethenyl-5-methoxy-1,4-benzoquinone. S. L. Xu, B. R. Yerxa, R. W. Sullivan, Y. Xiong, and H. W. Moore, School of Physical Sciences, University of California, Irvine, Irvine, CA 92717.

2692R*   [3+2]-Anionic Electrocyclization Using 2,3-Bis(phenylsulfonyl)-1,3-butadiene.
Z. Ni and A. Padwa, Department of Chemistry, Emory University, Atlanta, GA 30322.

2694   Cyclopentanone Annulation via Cyclopropanone Derivatives: cis-3a-Hydroxy-7a-methylbenzinden-1-one.
M. J. Bradlee and P. Helquist, Department of Chemistry and Biochemistry, University of Notre Dame, Notre Dame, IN 46556.

2696   Regio- and Stereoselective Carboxylation of Allylic Barium Reagents: (E)-4,8-Dimethyl-3,7-nonadienoic Acid.
A. Yanagisawa, K. Yasue, and H. Yamamoto, School of Engineering, Nagoya University, Chikusa, Nagoya 464-01, Japan.

2697R   Synthesis and [3+2] Cycloaddition of 2,2-Dialkoxy-1-methylene-cyclopropane: 5,5-Dimethyl-3,7-dioxa-1-methylenespiro[2.5]octane.
S. Yamago and E. Nakamura, Department of Chemistry, Tokyo Institute of Technology, Meguro, Tokyo 152, Japan.

2698R   General Synthesis of Substituted Cyclopropenones and Their Acetals; 2-Ethyl-2-cyclopropen-1-one Acetal and 2-(1-Hydroxyhexyl)-2-cyclopropen-1-one.
M. Isaka and E. Nakamura, Department of Chemistry, Tokyo Institute of Technology, Meguro, Tokyo 152, Japan.

2699   Synthesis of Epoxides Using Dimethyldioxirane.
R. W. Murray and Megh Singh, Department of Chemistry, University of Missouri-St. Louis, St. Louis, MO 63121.

2706*   Pyridine-Derived Triflating Reagents: N-(2-Pyridyl)triflimide and N-(5-Chloro-2-pyridyl)triflimide.
D. L. Comins, A. Dehghani, C. J. Foti, and S. P. Joseph, Department of Chemistry, North Carolina State University, Raleigh, NC 27695-8204.

2707   Photocatalyzed Addition of Methanol to (5S)-5-O-tert-Butyldimethyl-siloxymethyl-furan-2(5H)-one: Synthesis of (4R,5S)-4-Hydroxy-methyl-5-O-tert-butyldimethylsiloxymethyl-furan-2(5H)-one.
J. Mann and A. C. Weymouth-Wilson, Department of Chemistry, University of Reading, Whiteknights, P.O. Box 224, Reading, RG6 2AD, England.

2708   Mesitylenesulfonyl Hydrazine and (1α,2α,6β)-2,6-Dimethylcyclo-
        hexanecarbonitrile and (1α,2β,6α)-2,6-Dimethylcyclohexane-
        carbonitrile.
        J. R. Reid, R. F. Dufresne and J. J. Chapman, Lorillard Research
        Center, 420 English Street, Greensboro, NC 27420.

2709   N-Benzyl-2,3-azetidinedione.
        C. Behrens and L. A. Paquette, Evans Chemical Laboratories, The
        Ohio State University, Columbus, OH 43210.

2711   Nitroacetaldehyde Diethyl Acetal.
        V. Jäger and P. Poggendorf, Institut für Organische Chemie und
        Isotopenforschung der Universität Stuttgart, Pfaffenwaldring 55, D-
        70569 Stuttgart, Germany.

2712   Acetylenic Ethers from Alcohols and Their Reduction to Z- and E-Enol
        Ethers: Preparation of 1-Menthoxy-1-butyne from Menthol and
        Conversion to (Z)- and (E)-1-Menthoxy-1-butene.
        N. Kann, V. Bernardes, and A. E. Greene, LEDSS, Chimie Recherche,
        Université Joseph Fourier, BP 53X, 38041 Grenoble Cédex, France.

2718   Rhodium-Catalyzed Heterocycloaddition of a Diazomalonate and a
        Nitrile.
        J. S. Tullis and P. Helquist, Department of Chemistry and
        Biochemistry, University of Notre Dame, Notre Dame, IN 46556.

2719   Phenyl Vinyl Sulfide.
        D. S. Reno and R. J. Pariza, Process Development, Chemical and
        Agricultural Products Division, Abbott Laboratories, North Chicago, IL
        60064.

2720*   (R,R)-N,N'-Bis-(3,5-di-tert-butylsalicylidene)-1,2-cyclohexanediamino
        Manganese(III) Chloride, a Highly Enantioselective Epoxidation
        Catalyst.
        J. F. Larrow and E. N. Jacobsen, Department of Chemistry, Harvard
        University, Cambridge, MA 02138.

2721   Preparation of 3-Bromopropiolic Esters: tert-Butyl 3-Bromopropiolate.
        J. Leroy, Département de Chimie, Ecole Normale Supérieure, 24, rue
        Lhomond, F-75231 Paris Cedex 05, France.

2722*   Nickel-Catalyzed Geminal Dimethylation of Allylic Dithioacetals: (E)-
        1-Phenyl-3,3-dimethyl-1-butene.
        T.-M. Yuan and T.-Y. Luh, Department of Chemistry, National Taiwan
        University, Taipei, Taiwan 106, Republic of China.

2723*   A Simple and Convenient Method for the Preparation of (Z) or (E)-γ-
           Iodo Allylic Alcohols and (Z)-β-Iodo Acrolein.
           I. Marek, C. Meyer and J.-F. Normant, Laboratoire de Chimie des
           Organoéléments, Tour 44-45 E.2. 4, Place Jussieu. Boite 183. 75252
           Paris Cedex 05, France.

CUMULATIVE AUTHOR INDEX

FOR VOLUMES 70, 71, 72, AND 73

This index comprises the names of contributors to Volume **70, 71, 72,** and **73,** only. For authors to previous volumes, see either indices in Collective Volumes I through VIII or the single volume entitled *Organic Syntheses, Collective Volumes, I, II, III, IV, V, Cumulative Indices,* edited by R. L. Shriner and R. H. Shriner.

Ma, S., **72**, 112
Maddox, J. T., **73**, 116
Magriotis, P. A., **72**, 252
Mander, L. N., **70**, 256
Marquais, S., **72**, 135
Maruoka, K., **72**, 95
Maruyama, K., **71**, 118, 125
Matthews, D. P., **72**, 209, 216
Mattson, M. N., **70**, 177
Maw, G. N., **71**, 48
McCarthy, J. R., **72**, 209, 216
McDonald, F. E., **70**, 204
McKee, B. H., **70**, 47
Medich, J. R., **71**, 133, 140
Meister, P. G., **70**, 226
Mergelsberg, I., **73**, 159
Merino, P., **72**, 21
Meyers, A. I., **71**, 107;
    73, 215, 221, 246
Mikami, K., **71**, 14
Miller, R. F., **73**, 61, 134
Miyai, J., **73**, 73
Miyaura, N., **71**, 89
Modi, S. P., **72**, 125
Moriwake, T., **73**, 184
Mozaffari, A., **70**, 68
Mudryk, B., **72**, 173
Mühlemann, C., **71**, 200
Murahashi, S.-I., **70**, 265
Murphy, C. K., **73**, 159
Myers, A. G., **72**, 104
Myles, D. C., **70**, 231

Nakagawa, Y., **73**, 94
Nakai, T., **71**, 14
Nakamura, E., **73**, 123
Narisawa, S., **71**, 14
Naruta, Y., **71**, 118, 125
Negrete, G. R., **73**, 201
Negron, A., **70**, 169
Ni, Z.-J., **70**, 240
Nielsen, R. B., **71**, 77
Nikolaides, N., **72**. 246
Nishigaichi, Y., **71**, 118
Nissen, J. S., **73**, 110
Nowick, J. S., **73**, 61
Noyori, R., **71**, 1; **72**, 74
Nugent, W. A., **70**, 272

O'Connor, E. J., **70**, 177
Obrecht, J.-P., **71**, 200
Oda, M., **73**, 240, 253
Oglesby, R. C., **72**, 125
Ohkuma, T., **71**, 1
Ohta, **72**, T., 74
Oi, R., **73**, 1
Okabe, M., **71**, 83; **72**, 48
Okada, T., **73**, 253
Okazoe, T., **73**, 73
Ooi, T., **72**, 95
Oshima, K., **72**, 180; **73**, 73
Overman, L. E., **70**, 111; **71**, 56, 63

Panek, J. S., **70**, 79
Pansare, S. V., **70**, 1, 10
Paolini, J. P., **72**, 209, 216
Paquette, L. A., **70**, 226; **71**, 175, 181;
    72, 57; **73**, 36, 44
Park, J. M., **70**, 18
Patel, M., **70**, 79
Patois, C., **72**, 241; **73**, 152
Pfau, M., **70**, 35
Pfiffner, A., **72**, 116
Pikul, S., **71**, 22, 30
Polin, J., **73**, 262
Porter, J. R., **72**, 189
Posner, G. H., **73**, 231
Prakash, G. K. S., **72**, 232
Presnell, M., **73**, 110
Protopopova, M. N., **73**, 13

Ramaiah, P., **72**, 232
Rapoport, H,. **70**, 29; **71**, 220, 226
Reddy R. E., **71**, 146
Reddy, R. T., **73**, 159
Reed, J. N., **72**, 163
Reichelt, I., **71**, 189
Reissig, H.-U., **71**, 189
Revial, G., **70**, 35
Rieke, L. I., **72**, 147
Rieke, R. D., **72**, 147
Rishton, G. M., **71**, 56, 63
Roberts, F. E., **73**, 144
Romines, K. R., **70**, 93; **71**, 133, 140
Royer, J., **70**, 54
Russ, W. K., **73**, 144

This index comprises subject matter for Volumes **70**, **71**, **72**, and **73** only. For subjects in previous volumes, see either the indices in Collective Volumes I through VIII or the single volume entitled *Organic Syntheses, Collective Volumes I-VIII, Cumulative Indices*, edited by J. P. Freeman.

The index lists the names of compounds in two forms. The first is the name used commonly in procedures. The second is the systematic name according to **Chemical Abstracts** nomenclature. Both are usually accompanied by registry numbers in parentheses. Also included are general terms for classes of compounds, types of reactions, special apparatus, and unfamiliar methods.

Most chemicals used in the procedure will appear in the index as written in the text. There generally will be entries for all starting materials, reagents, intermediates, important by-products, and final products. Entries in capital letters indicate compounds appearing in the title of the preparation.

Acetamidoacrylic acid, **70**, 6

p-Acetamidobenzenesulfonyl azide: Sulfanilyl azide, N-acetyl-; Benzenesulfonyl

azide, 4-(acetylamino)-; (2158-14-7), **70**, 93; **73,** 16, 20, 24

p-Acetamidobenzenesulfonyl chloride: Benzenesulfonyl chloride, 4-(acetylamino)-;

(121-60-8), **70**, 93

Acetic acid; (64-19-7), **71**, 168

Acetic anhydride; (108-24-7), **70**, 69, 130; **71**, 72; **72**, 49; **73**, 35

Acetonitrile: TOXIC. (75-05-8), **71**, 227; **73**, 15, 23, 135, 136, 143

anhydrous, **70**, 2

1-ACETOXY-2-BUTYL-4-METHOXYNAPHTHALENE: 1-NAPHTHALENOL,

2-BUTYL-4-METHOXY-, ACETATE; (99107-52-5), **71**, 72

(4R)-(+)-Acetoxy-2-cyclopenten-1-one: 2-Cyclopenten-1-one, 4-(acetyloxy)-, (R)-;

(59995-48-1), **73**, 36, 37, 38, 43

(±)-cis-4-Acetoxy-1-hydroxycyclopent-2-ene: 4-Cyclopentene-1,3-diol, monoacetate,

cis-; (60410-18-6), **73**, 25, 27, 35

(1S,4R)-(+)-4-Acetoxy-1-hydroxycyclopent-2-ene: 4-Cyclopentene-1,3-diol,

monoacetate, (1S-cis); (60410-16-4), **73**, 25, 26, 35

4-Acetoxy-3-nitrohexane: 3-Hexanol, 4-nitro-, acetate; (3750-83-2), **70**, 69

(2S,3S)-3-ACETYL-8-CARBOETHOXY-2,3-DIMETHYL-1-OXA-8-

AZASPIRO[4.5]DECANE, **71**, 63

Acetyl chloride; (75-36-5), **71**, 148; **72**, 33; **73**, 185, 189, 200

Acetyl cholinesterase (from electric eel): Esterase, acetyl choline; (9000-81-1),

73, 26, 29, 32, 33, 35

Acetylenes, coupling reactions, **72**, 104

Acetylenes, perfluoroalkyl, **70**, 246

α-ACETYLENIC ESTERS, **70**, 246

Acetylferrocene: HIGHLY TOXIC: Ferrocene, acetyl-; (1271-55-2),

73, 262, 264, 267, 269

O-Acetylserine, **70**, 6

Acrolein: 2-Propenal; (107-02-8), **71**, 236; **72**, 190; **73**, 2, 12

Acryloyl-2-oxazolidinone: 2-Oxazolidinone, 3-acryloyl-; (2043-21-2), **71**, 31

Activated magnesium, **73**, 52

N-Acylaziridines, **70**, 106

α-Acylmethylenephosphoranes, **70**, 251

(ACYLOXY)BORANE COMPLEXES, CHIRAL, **72**, 86

3-ACYLTETRAHYDROFURANS, Stereocontrolled preparation, **71**, 63

1-ADAMANTYL CALCIUM HALIDES, **72**, 147

1-(1-ADAMANTYL)CYCLOHEXANOL: CYCLOHEXANOL,

1-TRICYCLO[3.3.1.1^{3,7}]DEC-1-YL-; (84213-80-9), **72**, 147

L-Alanine; (56-41-7), **71**, 227

Aliquat 336: Ammonium, methyltrioctyl-, chloride; 1-Octanaminium, N-methyl-N,N-

dioctyl-, chloride; (5137-55-3), **73**, 145, 147, 151, 162, 166, 158, 173

ALKENES, SYNTHESIS, **73**, 61

ALKENYLCHROMIUM REAGENTS, **72**, 180

Z-Alkenyl ethers, **73**, 78

α-Alkoxyorganolithiums, **71**, 143, 144

α-Alkoxyorganostannanes, **71**, 143

α-Alkylated piperidines, **70**, 57

Alkylation-metalation of a trialkyl phosphate, **72**, 244

9-ALKYL-9-BORABICYCLO[3.3.1]NONANES, **71**, 89

Alkylzinc iodide, **70**, 201

ALKYNE-IMINIUM ION CYCLIZATIONS, **70**, 111

Alkynyl(phenyl)iodonium sulfonates, **70**, 223

α-Aminonitrile, **70**, 57

(S)-3-AMINO-2-OXETANONE p-TOLUENESULFONATE: 2-OXETANONE, 3-AMINO-, (S)-, 4-METHYLBENZENESULFONATE; (112839-95-9), **70**, 10

Amipurimycin studies, **70**, 27

Ammonia; (7664-41-7), **72**, 253; **73**, 87, 175, 177, 178, 183, 223

Ammonium formate: Formic acid, ammonium salt; (540-69-2), **72**, 128

Ammonium hydroxide; (1336-21-6), **73**, 87, 271, 277

Annulated lactams, **70**, 106

Antimony trichloride, **72**, 210

Arsoles, **70**, 276

Ascorbic acid, **72**, 3

Asparagine monohydrate: L-Asparagine; (70-47-3), **73**, 202, 205, 213

Asymmetric catalysis, **71**, 1, 14, 30

ASYMMETRIC DIELS-ALDER REACTION, **72**, 86

Asymmetric dihydroxylation, **70**, 51; **73**, 8

ASYMMETRIC HYDROGENATION, **72**, 74

Asymmetric hydroxylation, **73**, 168, 170

ASYMMETRIC SYNTHESIS OF SUBSTITUTED PIPERIDINES, **70**, 54

Azetidinones, **72**, 19

Azides, sulfonyl, **70**, 97

β-Azidoalanine, **70**, 15

Azido lactams, **70**, 106

Azidotrimethylsilane: HIGHLY TOXIC: Silane, azidotrimethyl-; (4648-54-8), **73**, 187, 191, 195, 196, 200

Azobisisobutyronitrile: Propanenitrile, 2,2'-azobis[2-methyl-; (78-67-1), **70**, 151, 167; **72**, 218

301

4,5-DIBENZOYL-1,3-DITHIOLE-1-THIONE: BENZENECARBOTHIOIC ACID,

S,S'-(2-THIOXO-1,3-DITHIOLE-4,5-DIYL) ESTER; (68494-08-6), **73**, 270, 272

(-)-Dibenzyl tartrate: Tartaric acid, dibenzyl ester, (+)- ; Butanedioic acid,

2,3-dihydroxy-, [R-(R*,R*)]-, bis(phenylmethyl) ester; (622-00-4), **72**, 87

Diborane-THF: Furan, tetrahydro-, compd. with borane (1:1); (14044-65-6), **70**, 131;

71, 207; **72**, 88, 202; **73**, 122

Dibromodifluoromethane: Methane, dibromodifluoro-; (75-61-6), **72**, 225

(1,3-Dibromo-3,3-difluoropropyl)trimethylsilane: Silane, (1,3-dibromo-3,3-

difluoropropyl)trimethyl-; (671-80-7), **72**, 225

3,5-Dibromo-5,6-dihydro-2(H)-pyran-2-one: 2H-Pyran-2-one, 3,5-dibromo-5,6-

dihydro-, (±)-; (56207-18-2), **73**, 232, 233, 235, 238

1,3-Dibromo-5,5-dimethylhydantoin: 2,4-Imidazolidinedione, 1,3-dibromo-5,5-

dimethyl-; (77-48-5), **70**, 151

1,2-Dibromoethane: Ethane, 1,2-dibromo-; (106-93-4), **70**, 196; **72**, 191, 254

Dibromomethane: Methane, dibromo-; (74-95-3), **71**, 147; **73**, 74, 76, 84

Dibromomethyllithium, **71**, 146; **73**, 73

1,1-Dibromopentane: Pentane, 1,1-dibromo-; (13320-56-4), **73**, 73, 74, 75, 84

Di-tert-butyl dicarbonate: Formic acid, oxydi-, di-tert-butyl ester; Dicarbonic acid,

bis(1,1-dimethylethyl) ester; (24424-99-5), **70**, 18, 140, 152; **71**, 202;

73, 86, 88, 93, 188, 192, 196, 198, 200

4,4'-Di-tert-butylbiphenyl: Biphenyl, 4,4'-di-tert-butyl-; 1,1'-Biphenyl, 4,4'-

bis(1,1-dimethylethyl)-; (1625-91-8), **72**, 173

2,6-Di-tert-butylphenol: Phenol, 2,6-di-tert-butyl-; Phenol, 2,6-bis

(1,1-dimethylethyl)-; (128-39-2), **72**, 95

Dibutyl telluride: Butane, 1,1'-tellurobis-; (38788-38-4), **72**, 155

Dichlorobis(triphenylphosphine)nickel: Nickel, dichlorobis(triphenylphosphine)-;

(14264-16-5), **70**, 241

309

311

317

3-Methyl-2-butenyl acetoacetate, **73**, 15

3-Methyl-2-butenyl diazoacetate, **73**, 16

Methyl chloroformate: Formic acid, chloro-, methyl ester; Carbonochloridic acid,

    methyl ester; (79-22-1), **73**, 202, 205, 213

16α-METHYLCORTEXOLONE: PREGN-4-ENE-3,20-DIONE, 17,21-DIHYDROXY-16-

    METHYL-, (16α)-(±)-; (122405-63-4), **73**, 123, 124, 126

Methyl cyanoformate: Carbonocyanidic acid, methyl ester; (17640-15-2), **70**, 256

2-Methylcyclohexanone: Cyclohexanone, 2-methyl; (583-60-8), **70**, 35

3-Methyl-2-cyclohexen-1-one, **70**, 261

2-METHYL-1,3-CYCLOPENTANEDIONE: 1,3-CYCLOPENTANEDIONE, 2-METHYL-;

    (765-69-5), **70**, 226

Methyl diazoacetate: Acetic acid, diazo-, methyl ester; (6832-16-2), **71**, 190

METHYL 3,3-DIMETHYL-4-OXOBUTANOATE: BUTANOIC ACID, 3,3-

    DIMETHYL-4-OXO-, METHYL ESTER; (52398-45-5), **71**, 189

Methyl 2,2-dimethyl-3-(trimethylsiloxy)-1-cyclopropanecarboxylate:

    Cyclopropanecarboxylic acid, 2,2-dimethyl-3-[(trimethylsilyl)oxy]-, methyl ester;

    (77903-45-8), **71**, 190

6-METHYL-6-DODECENE, **73**, 50, 52

(E)-6-Methyl-6-dodecene: 6-Dodecene, 6-methyl, (E)-; (101146-61-6), **73**, 60

(Z)-6-Methyl-6-dodecene: 6-Dodecene, 6-methyl-, (Z)-; (101165-44-0), **73**, 60

2-Methylene-1,3-dithiolane: 1,2-Dithiolane, 2-methylene-; (26728-22-3), **71**, 175, 182

Methylene transfer reagent, **70**, 187

Methyl glyoxylate: Glyoxylic acid, methyl ester; Acetic acid, oxo-, methyl ester;

    (922-68-9), **71**, 15

N-Methylhydrazine: Hydrazine, methyl-; (60-34-4), **72**, 16

Methyl sulfide: Methane, thiobis-; (75-18-3), **71**, 214

6-METHYL-2,3,4,5-TETRAHYDROPYRIDINE N-OXIDE: PYRIDINE, 2,3,4,5-
TETRAHYDRO-6-METHYL-, 1-OXIDE; (55386-67-9), **70**, 265

Methyl trifluoromethanesulfonate: Methanesulfonic acid, trifluoro-, methyl ester;
(333-27-7), **71**, 110

(Z)-16α-Methyl-20-trimethylsiloxy-4,17(20)-pregnadien-3-one: Pregna-4,17(20)-dien-
3-one, 16-methyl-20-[(trimethylsilyl)oxy]-, (16α,17Z)-(±)-; (122315-01-9),
73, 124, 132

2-Methyl-1-(trimethylsiloxy)propene: Silane, trimethyl[(2-methyl-1-propenyl)oxy]-;
(6651-34-9), **71**, 189

Methyl vinyl ketone, **70**, 35

Michael-type alkylation, **70**, 40

Mitsunobu conditions, **70**, 6

MITSUNOBU INVERSION OF STERICALLY HINDERED ALCOHOLS, **73**, 110, 113

Molybdophosphoric acid, **70**, 31

Monobactam antibiotics, **72**, 19

Mono(2,6-dimethoxybenzoyl)tartaric acid: Butanedioic acid, 2-[(2,6-dimethoxy-
benzoyl)oxy-3-hydroxy-, [R-(R*,R*)]-; (116212-44-3), **72**, 87

Mosher ester, **70**, 24, 48; **73**, 7, 30

NAPHTHALENEDIOLS, 2-SUBSTITUTED, SYNTHESIS, **71**, 72

1-Naphthalenesulfonyl azide, **70**, 97

(R)-1-(1-Naphthyl)ethylamine: 1-Naphthalenemethylamine, α-methyl-, (R)-(+)-;
1-Naphthalenemethanamine, α-methyl-, (R)-; (3886-70-2), **72**, 79

Nerol, **72**, 82

Nickel acetylacetonate: Nickel, bis(2,4-pentanedionato)-; Nickel,
bis(2,4-pentanedionato-O,O')-, (SP-4-1)-; (3264-82-2), **71**, 84

p-Phenylbenzoyl chloride: 4-Biphenylcarbonyl chloride; [1,1'-Biphenyl]-4-carbonyl

    chloride; (14002-51-8), **72**, 53

trans-4-Phenyl-3-buten-2-one: 3-Buten-2-one, 4-phenyl-, (E)-; (1896-62-4),

    73, 134, 136, 143

2-Phenylcyclohexanone: Cyclohexanone, 2-phenyl-; (1444-65-1), **72**, 217

S-Phenyl decanoate: Decanethioic acid, S-phenyl ester; (51892-25-2),

    73, 61, 53, 72

Phenyldimethylsilyl chloride: Aldrich: Chlorodimethylphenylsilane: Silane,

    chlorodimethylphenyl-; (768-33-2), **73**, 51, 55, 59

(E)-5-PHENYLDODEC-5-EN-7-YNE: BENZENE, (1-BUTYL-1-OCTEN-3-YNYL)-, (E)-;

    (111525-79-2), **70**, 215

2-(2-Phenylethenyl)-1,3-dithiolane: 1,3-Dithiolane, 2-(2-phenylethenyl)-; (5616-58-0),

    70, 240

9-Phenyl-9-fluorenol: Fluoren-9-ol, 9-phenyl-; 9H-Fluoren-9-ol, 9-phenyl-;

    (25603-67-2), **71**, 220, 228

(S)-N-(9-Phenylfluoren-9-yl)alanine: L-Alanine, N-(9-phenyl-9H-fluoren-9-yl)-;

    (105519-71-9), **71**, 226

(-)-Phenylglycine, **70**, 55

(S)-(+)-2-Phenylglycine: Glycine, 2-phenyl-, L-; Benzeneacetic acid, α-amino-,

    (S)-; (2935-35-3), **73**, 224, 230

(S)-(+)-2-Phenylglycinol: Phenethyl alcohol, β-amino-, L-; (20989-17-7),

    73, 222, 224, 227, 229

(-)-Phenylglycinol: Benzeneethanol, β-amino-, (R)-; (56613-80-0), **70**, 54

Phenyllithium: Lithium, phenyl-; (591-51-5), **71**, 223

Phenylmagnesium bromide: Magnesium, bromophenyl-; (100-58-3), **72**, 32

341